eTELEMED 2017

The Ninth International Conference on eHealth, Telemedicine, and Social Medicine

MATH 2017

The International Symposium on Mobile and Assistive Technology for Healthcare

March 19 – 23, 2017

Nice, France

eTELEMED 2017 Editors

Marike Hettinga, Windesheim University of Applied Sciences, the Netherlands
Yoshitoshi Murata, Iwate Prefectural University, Japan
Britt Östlund, Royal Institute of Technology (KTH), Sweden
Bobby Gheorghiu, Canada Health Infoway/Inforoute Santé du Canada, Canada
SeungHee Lee, University of Tsukuba, Japan
José Machado, University of Minho, Portugal
Carlos Filipe da Silva Portela, University of Minho, Portugal
Aranka Dol, Hanzehogeschool University of Applied Sciences, the Netherlands
Paolo Perego, Politecnico di Milano, Italy

Printed from e-media with permission by:

Curran Associates, Inc.
57 Morehouse Lane
Red Hook, NY 12571

Some format issues inherent in the e-media version may also appear in this print version.

Printed by Curran Associates, Inc. (2017)

International Academy, Research, and Industry Association (IARIA)
412 Derby Way
Wilmington, DE 19810

Phone: (408) 893-6407
Fax: (408) 527-6351

petre@iaria.org

Additional copies of this publication are available from:

Curran Associates, Inc.
57 Morehouse Lane
Red Hook, NY 12571 USA
Phone: 845-758-0400
Fax: 845-758-2633
Email: curran@proceedings.com
Web: www.proceedings.com

Table of Contents

eTELEMED 2017

Forward

The ninth edition of The International Conference on eHealth, Telemedicine, and Social Medicine (eTELEMED 2017), held in Nice, France, March 19 - 23, 2017, considered advances in techniques, services, and applications dedicated to a global approach of eHealth.

Development of wireless homecare, of special types of communications with patient data, of videoconferencing and telepresence, and the progress in image processing and date protection increased the eHealth applications and services, and extended Internet-based patient coverage areas. Social and economic aspects as well as the integration of classical systems with the telemedicine systems are still challenging issues.

eTELEMED 2017 provided a forum where researchers were able to present recent research results and new research problems and directions related to them. The topics covered aspects from classical medicine and eHealth integration, systems and communication, devices, and applications.

eTELEMED 2017 also featured the following Symposium:

- MATH 2017: The International Symposium on Mobile and Assistive Technology for Healthcare

We take this opportunity to thank all the members of the eTELEMED 2017 Technical Program Committee as well as the numerous reviewers. The creation of such a broad and high-quality conference program would not have been possible without their involvement. We also kindly thank all the authors who dedicated much of their time and efforts to contribute to the eTELEMED 2017. We truly believe that, thanks to all these efforts, the final conference program consists of top quality contributions.

This event could also not have been a reality without the support of many individuals, organizations, and sponsors. We are grateful to the members of the eTELEMED 2017 organizing committee for their help in handling the logistics and for their work to make this professional meeting a success.

We hope that eTELEMED 2017 was a successful international forum for the exchange of ideas and results between academia and industry and for the promotion of progress in eHealth and Telemedicine research.

We also hope that Nice provided a pleasant environment during the conference and everyone saved some time for exploring this beautiful city.

eTELEMED 2017 Chairs

eTELEMED Steering Committee
Lisette Van Gemert-Pijnen, University of Twente - Enschede, the Netherlands
Marike Hettinga, Windesheim University of Applied Sciences, the Netherlands
Joseph Finkelstein, Center for Bioinformatics and Health Analytics - Columbia University, USA
Åsa Smedberg, Stockholm University, Sweden
Yoshitoshi Murata, Iwate Prefectural University, Japan
Anthony Glascock, Drexel University, USA
Britt Östlund, Royal Institute of Technology (KTH), Sweden

eTELEMED Industry/Research Advisory Committee
Anne G. Ekeland, Norwegian Centre for e-Health Research, Norway
Salah H. Mandil, eStrategies & eHealth for WHO and ITU - Geneva, Switzerland
Bobby Gheorghiu, Canada Health Infoway/Inforoute Santé du Canada, Canada

MATH 2017 Chairs
Paolo Perego, Politecnico di Milano, Italy
Cristiana Degano, Gruppo Sigla, Italy
Maurizio Caon, HES.SO, Switzerland
SeungHee Lee, University of Tsukuba, Japan

eTELEMED 2017

COMMITTEE

eTELEMED Steering Committee

Lisette Van Gemert-Pijnen, University of Twente - Enschede, the Netherlands
Marike Hettinga, Windesheim University of Applied Sciences, the Netherlands
Joseph Finkelstein, Center for Bioinformatics and Health Analytics - Columbia University, USA
Åsa Smedberg, Stockholm University, Sweden
Yoshitoshi Murata, Iwate Prefectural University, Japan
Anthony Glascock, Drexel University, USA
Britt Östlund, Royal Institute of Technology (KTH), Sweden

eTELEMED Industry/Research Advisory Committee

Anne G. Ekeland, Norwegian Centre for e-Health Research, Norway
Salah H. Mandil, eStrategies & eHealth for WHO and ITU - Geneva, Switzerland
Bobby Gheorghiu, Canada Health Infoway/Inforoute Santé du Canada, Canada

eTELEMED 2017 Technical Program Committee

James G. Anderson, Purdue University, USA
Adina Astilean, Technical University of Cluj-Napoca, Romania
Luigi Benedicenti, University of Regina, Canada
Andrzej Cacko, Medical University of Warsaw, Poland
Marcos C. d'Ornellas, Universidade Federal de Santa Maria, Brazil
Aranka Do, Hanze University of Applied Sciences, Netherlands
Georg Duftschmid, Medical University Vienna, Austria
Leila Eadie, University of Aberdeen, UK
Øystein Eiring, National Institute of Public Health, Norway
Anne G. Ekeland, Norwegian Centre for e-Health Research, Norway
Henrik Eriksson, Linköping University, Sweden
Ivan Evgeniev, Technical University of Sofia, Bulgaria
Manuel Filipe Santos, Universidade do Minho | Research Centre Algoritmi, Portugal
Joseph Finkelstein, Center for Bioinformatics and Health Analytics - Columbia University, USA
Audrey Dunn Galvin, University College Cork, UK
Catherine Garbay, Centre National de la Recherche Scientifique - Laboratoire d'Informatique de Grenoble, France
Anthony Gelibert, LCIS, France
Gersende Georg, Haute Autorité de Santé, France
Bobby Gheorghiu , Canada Health Infoway/Inforoute Santé du Canada, Canada
Piero Giacomelli, SPAC SPA - Arzignano (Vicenza), Italy
Anthony Glascock, Drexel University, USA
Wojciech Glinkowski, Polish Telemedicine Society / Center of Excellence "TeleOrto", Poland
Conceição Granja, Norwegian Centre for eHealth Research | University Hospital of North Norway, Norway
David Greenhalgh, University of Strathclyde, Scotland
Päivi Hämäläinen, National Institute for Health and Welfare, Finland

MATH 2017 Chairs

Paolo Perego, Politecnico di Milano, Italy
Cristiana Degano, Gruppo Sigla, Italy
Maurizio Caon, HES.SO, Switzerland
SeungHee Lee, University of Tsukuba, Japan

Program Committee Members

Abdulellah A. Alaboudi, Staffordshire University, United Kingdom
Sibylle Artz, University of Victoria, Canada
Hao Cai, Aalborg University, Denmark, & Guangdong University of Technology, China
Maurizio Caon, HES.SO, Switzerland
Carme Carrion, Universitat Oberta de Catalunya, Spain
Cristiana Degano, Gruppo Sigla, Italy
Yannis Ellul, University of Patras (Clinical University), Greece
Carlo Emilio Standoli, Politecnico di Milano, Italy
Florian Evéquoz, HES-SO and University of Fribourg, Switzerland
SeungHee Lee, University of Tsukuba, Japan
Inge Lissau, Skane University Hospital, Sweden
Fulvio Marchetti, Healthy Reply, Italy
Vasileios Megalooikonomou, University of Patras, Greece
Mauro Migliardi, University of Padova, Italy
Franco Molteni, Ospedale Villa Beretta, Italy
Giandomenico Nollo, University of Trento, Italy
Paolo Perego, Politecnico di Milano, Italy
Giovanni Saggio, University of Rome Tor Vergata, Italy
Ziba Vaghri, University of Victoria, Canada
Nadine Vigouroux, Université Paul Sabatier, France

Copyright Information

The Problem of Retrieving Clinical Guidelines in Self-Care

Juha Puustjärvi
Innomedi
Helsinki, Finland
email: juha.puustjarvi@cs.helsinki.fi

Leena Puustjärvi
The Pharmacy of Kaivopuisto
Helsinki, Finland
email: leena.puustjarvi@kolumbus.fi

***Abstract*— Self-care is what people do for themselves to establish and maintain health, prevent and deal with illness. A clinical guideline is a document with the aim of guiding decisions and criteria regarding diagnosis, management, and treatment in specific areas of healthcare. Clinical guidelines have been in use for thousands of years during the entire history of medicine. An interesting question arising from these notions is whether clinical guidelines can be used in self-care to ensure that the treatment is in line with the medical recommends. A problem, however, is that the retrieval methods of clinical guidelines are mainly aimed at healthcare personnel who are taught to retrieve relevant guidelines. In order to simplify patients' tasks in retrieving guidelines, we have designed an ontology (vocabulary) for annotating clinical guidelines. A patient can then query clinical guidelines by keyword expressions, which are intuitive and clear for patients. Technically annotations are presented according to the Semantic web technologies, and so they are machine understandable. Further, guidelines can be retrieved from a variety of sources within one query. In addition, the system can be easily extended such that guidelines and other relevant health-oriented information can be retrieved in a query.**

Keywords – Self-Care; Clinical Guidelines; Evidence-Based Medicine; Semantic Web; SPARQL

I. INTRODUCTION

Nowadays, health care provision is moving towards self-care from the physician centric model where the treatment decisions are made almost exclusive by physicians. This is a challenging trend as it is estimated that 70% to 95% of all illnesses could be managed without the intervention of a doctor [1], i.e., by self-care [2].

To promote self-care some Personal Health Records (PHRs) already provide links to websites that provide information concerning patient's treatment and medication. However, many patients have regarded these sites to be overly commercial, and often patients cannot determine the source of the published medical information [3]. In addition, browsing these pages has turned out to be long lasting and frustrating as their provided medical information is not relevant, is overly specific for patients or is overly superficial [4].

After all, success in self-care requires patients to have relevant medical treatment information [5]. Traditionally, such information is provided for physicians in clinical guidelines, which are document with the aim of guiding decisions and criteria regarding diagnosis, management, and treatment in specific areas of healthcare [6].

Clinical guidelines are based on an examination of current evidence within the paradigm of *evidence-based medicine* (EBM) [7]. It emphasizes the use of evidence from well designed and conducted research in healthcare decision-making [8]. EBM is one of the most significant developments in the clinical use of information over the last decades [9]. Practicing EBM represents a thorough change in the way that physicians are taught to think and retrieve relevant guidelines [10].

There are a variety of ways how the digital volumes of guidelines can be organized. For example, guidelines may be grouped in an alphabetic order of the disorders they deal with, grouped by specialities, or grouped by body parts. Accessing guidelines from such digital volumes have proven to be easy for physicians and healthcare professionals as healthcare providers are obliged to know how to find relevant guidelines.

Instead searching relevant guidelines from such volumes is not necessary a trivial task for patients. This is regrettable as clinical guidelines are intended also for patients' use. Furthermore, clinical guidelines have special versions for different user groups.

Our argument is that clinical guideline oriented services should also provide specific retrieval methods for patients, as they are not taught to think in accordance with EBM. For example, a pregnant patient may be interested to know whether she has some risks in using pain drugs. In this case, finding relevant information from guidelines dealing with pregnancy and from guidelines dealing with pain drugs may be long lasting and frustrating process.

In order to develop appropriate retrieval methods for patients, we have annotated clinical guidelines by appropriate metadata items such as keywords and classification identifiers. A patient can then query clinical guidelines by Boolean expressions [11] comprising of operands and operations. The operands are the used keywords and the operands are typically "and", "or", and "not". The Boolean model is intuitive and clear. For example, the query "pregnancy and pain drugs" returns the guidelines (if any) that are annotated by these keywords.

Another problem concerning the retrieval of clinical guidelines is that they are fragmented. That is, there are many sites publishing guidelines, and therefore searching relevant guidelines from a variety of sites is often hard and long lasting. In order to avoid such weakness, a key point in our solution is that in one query patient can retrieve guidelines located in many web sites. This, however, requires that all the sites use the same vocabulary (ontology) and technology in annotating clinical guidelines.

Technically, in our solution, each site is assumed to provide a web service that supports SPARQL queries. SPARQL is a query language developed for Semantic web [12].

The rest of the paper is organized as follows. First, in Section II, we shortly present the architecture of the system that we are developing. Then, in Section III, we consider the Guideline ontology that we use in annotating guidelines. In Section IV, we give an example of a SPARQL query that exploits the Guideline ontology in searching guidelines from two data stores. In Section V, we consider the extension of the system by medicinal data. Finally, Section VI concludes the paper by considering our future research.

II. THE ARCHITECTURE OF THE GUIDELINE SERVER

In our developed architecture, users (patients) retrieve clinical guidelines through the Clinical Guideline Portal (Fig. 1). The portal provides a set of applications each providing a specific search method, such as searches based on keywords or body parts. These applications communicate with the SPARQL Processor, which access clinical guidelines through their Web Services.

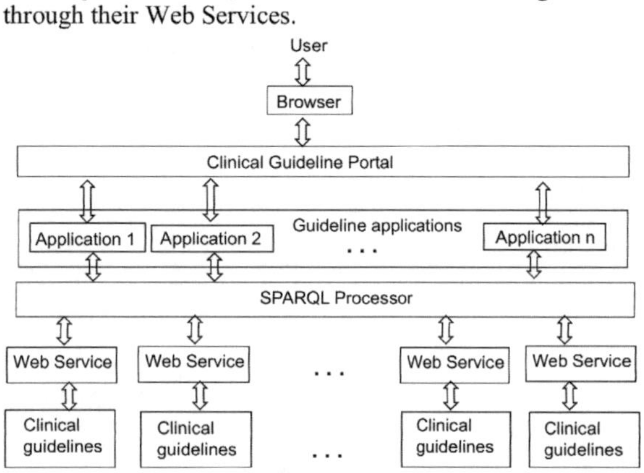

Figure 1. The architecture of the Clinical Guideline Retrieval system.

A useful feature of the SPARQL Processor is that it is capable for querying many RDF-formatted [13] data sources within one SPARQL query.

III. GUIDELINE ONTOLOGY

An ontology characterizes the meaning of concepts and their relationships [14. Within computer science, an ontology is usually represented by using classes, properties, and their values as modeling primitives [15]. Hence an ontology provides a systematic way to standardize the used metadata items [16]. As an example consider our defined Guideline Ontology, which is graphically presented in Fig. 2.

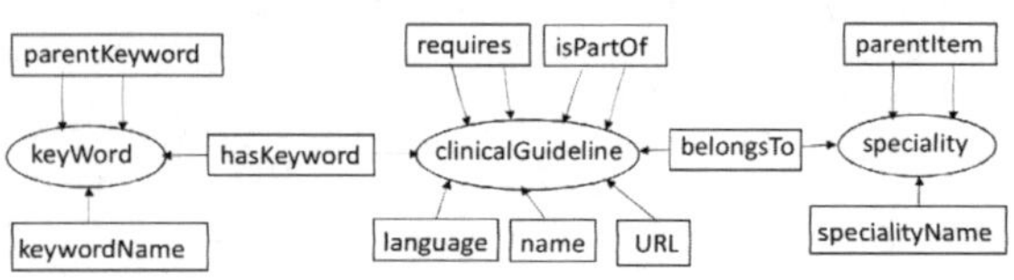

Figure 2. Guideline Ontology.

In this figure, ellipses represent classes and rectangles represent data type as well as object properties. These representations are modeling primitives in OWL [17]. Object properties (e.g., belongsTo) relate objects to other objects (or to itself, e.g., requires), and datatype properties (e.g., specialityName) relate objects to datatype values.

An instance of the Guideline Ontology is presented in RDF in Fig. 3. RDF is a key for representing machine understandable data. It is a data model with a variety of syntaxes for storing data files. By RDF we can express facts with tree-part statements called as triples. The subject identifies the thing being described, predicate is a property name, and object is property value. That is, each triple is like a little sentence that states a fact. However, RDF in itself does not bring interoperable semantics. It depends on the expression power of the used vocabulary. By a vocabulary we refer to a set of ontologies, which specifies the used terms and their semantics.

In order to illustrate the dependency of RDF and ontologies consider the RDF-description (a set of RDF-statements concerning the same subject) of Fig. 3. The RDF-description states, by using the Guideline Ontology as a vocabulary, that a guideline named *Pregnancy* has keywords *pregnancy* and pain *drugs*, and the guideline is a part of a guideline named *gynaecological diseases*. That is, pregnancy is a subclassification of gynecological diseases.

```
<rdf:RDF
  xmlns : rdf="http://www.w3.org/1999/02/22-rdf-syntax-ns#"
  xmlns : no=http://www.helsinki.fi/GuidelineOntology#>
  <rdf:Description rdf:about="pregnancyGuideline">
    <rdf:type rdf:resource="&po;clinicalGuideline"/>
    <no : hasKeyword>pregnancy</no :hasKeyword>
    <no : hasKeyword>pian_drugs</no :hasKeyword>
    <no : isPartOf>gynaecological diseases</no:isPartOf>
  </rdf : Description>
</rdf:RDF>
```

Figure 3. An annotation of a guideline.

The Guideline Ontology enables to attach parenKeyword to each keyword meaning that keywords comprise taxonomies (hierarchies). The logic behind taxonomy is that when one goes up the taxonomy toward the root, the keywords become more general, and respectively when one goes down towards the leaves the keywords become more specialized [16]. We can also state this in a more formal way: depending on the direction of the link each link between a parent and a child node represents a subclassification relation or superclassification relation. For example, type 2 diabetes is a subclassification of diabetes, and diabetes is a superclassification of type 2 diabetes.

Taxonomies can be exploited in searching, if the used keywords return too many guidelines. For example, we could replace keyword *diabetes* by its subclassification *type 1 diabetes*, *type 2 diabetes*, or *gestational diabetes* (a condition in which women without previously diagnosed diabetes exhibit high blood glucose levels during pregnancy).

IV. GUERYING GUIDELINES FROM MULTIPLE DATASTORES BY SPARQL

SPARQL is a query language that is able to retrieve and manipulate data stored in RDF-format. It was made a standard by the RDF Data Access Working Group of the World Wide Web Consortium, and is recognized as one of the key technologies of the semantic web [18].

A SPARQL query specifies the pieces of data that meets the stated requirements. The requirements are described with triple patterns, which are similar to RDF triples but they include variables to add flexibility in how they match against the queried data [18].

There is a variety of SPARQL technologies available for running queries against data both locally and remotely. For example, SPARQL provides two ways for querying remotely: using FROM keyword or using SERVICE keyword. In the former way the FROM keyword names a dataset to query that may be local or remote file. In the latter and our used way, instead of pointing at an RDF file somewhere, a SPARQL endpoint is pointed. An *endpoint* is a Web service that accepts SPARQL queries, runs the queries, and finally returns the result.

Further, Federated Queries in SPARQL allow searching multiple datasets with one query. For each dataset is created a subquery which access datasets by using SERVICE keywords. To illustrate this, consider the federated SPARQL query presented in Fig. 4. It accesses two datasets through SPARQL endpoints. The result of the query is the union of the results of the two subqueries. The query returns links (URLs) to those clinical guidelines that are annotated by the keywords *pain drug* and *pregnant*. The query is based on the ontology presented in Fig. 2 (the prefix *no* in the query refers to this ontology).

```
PREFIX owl: <http://www.w3.org/2002/07/owl#>
PREFIX rdf: <http://www.w3.org/1999/02/22-rdf-syntax-ns#>
PREFIX no: <http://www.cs.helsinki.fi/registryOntology#>
SELECT ?URL
WHERE
{
 SERVICE <http://documentRegistry_A/sparql>
 { SELECT ?URL
  WHERE
  {
   ?no:name no:hasKeyword pain_drug ;
   ?no:name no:hasKeyword pregnant ;
   ?no:name no:URL ?URL .
  }
 }
 SERVICE <http://documentRegistry_B/sparql>
 { SELECT ?URL
  WHERE
  {
   ?no:name no:hasKeyword pain_drug ;
   ?no:name no:hasKeyword pregnant ;
   ?no:name no:URL ?URL .
  }
 }
}
```

Figure 4. A simple federated SPARQL query.

V. EXTENDING THE GUIDELINE SERVER

The Clinical Guideline Retrieval System presented in Section II is focused only on retrieving relevant guidelines. However there are also websites that have information about medication, and thus they can provide valuable supplementary information for self-care. For example, a patient may be interested to know more about a pain drug recommend in a clinical guideline.

A problem however is that most health-oriented web sites are developed only for human consumption, and so they as well as their metadata is not machine understandable. In order to transform their metadata in machine understandable format, we have developed an appropriate ontology (called Medicine Ontology) for annotating the documents that deal with medicine.

A portion of the Medicine Ontology is graphically presented in Fig. 5. In this graphical representation ellipses represent classes and rectangles represent data type and object properties. Object properties relate objects to other objects while data type properties relate objects to datatype values. Classes, data type properties and object properties are modelling primitives in OWL.

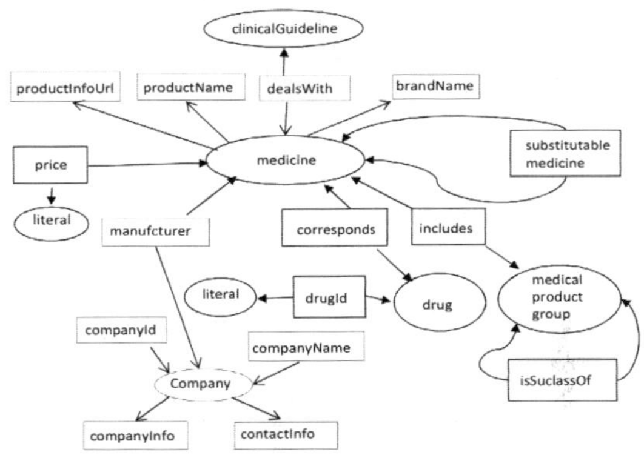

Figure 5. Graphical presentation of the Medicine Ontology.

Note that the Medicine Ontology and the Guideline Ontology are overlapping in the sense that they both include the class clinicalGuideline. As a result we can use the Medicine Ontology in annotating clinical guidelines by medicinal data. For example, in Fig. 6, the annotation presented in Fig. 3 is extended by stating that the instance deals with Aspirin.

```
<rdf:RDF
  xmlns : rdf="http://www.w3.org/1999/02/22-rdf-syntax-ns#"
  xmlns : no=http://www.helsinki.fi/GuidelineOntology#>
  xmlns : ro=http://www.helsinki.fi/MedicineOntology#>
  <rdf:Description rdf:about="pregnancyGuideline">
    <rdf:type rdf:resource="&po;clinicalGuideline"/>
    <no : hasKeyword>pregnancy</no :hasKeyword>
    <no : hasKeyword>pian_drugs</no :hasKeyword>
    <no : isPartOf>gynaecological diseases</no:isPartOf>
    <ro : dealsWith>Aspirin</ro :hasKeyword>
  </rdf : Description>
</rdf:RDF>
```

Figure 6. Using two ontologies in annotating a clinical guideline.

Similar to the extension on Medicine ontology we can also introduce other ontologies as well. For example, introducing an ontology for the terms of welfare might also be useful.

VI. CONCLUSION

Web-based e-health models and the Semantic Web support each other: Web-based health-oriented services provide a new paradigm for sharing health information while Semantic Web enables the presentation of information in a machine understandable form. As a result the use of Web-based e-health services is rapidly increasing. Yet, by using together a variety of services we can still achieve new services that would not be achievable by independently working systems. For example, we can achieve new outcomes by the interoperation of the systems developed for self-care, welfare, and smart homes. Technically the interoperation can be implemented by a SPARQL processor.

So far we have used these technologies in developing a patient-friendly retrieve method for clinical guidelines. We have also extended this solution by the services that provide information about medicines. In our future work we will consider how the information of clinical guidelines, medication, welfare and smart home can be used together in developing new services.

REFERENCES

[1] "The Role of the Pharmacist in Self-Care and Self-Medication". Available at at: http://apps.who.int/medicinedocs/pdf/whozip32e/whozip32e.pdf. Retrieved: 6, 2016.

[2] "What is self-care?" ISF International Self-Care Foundation. Available at: http://isfglobal.org/what-is-self-care/. Retrieved: 7, 2016.

[3] "Health Education in Self-Care: Possibilities and Limitations. Report of a Scientific Consultation". Geneva,Switzerland: World Health Organization; November 21–25, 1983.

[4] S. Lowell, S. Levin, and L. Idler, "Self-Care in Health".Vol. 4: 181-201, 1983.

[5] "Self-care in the Context of Primary Health Care". Report of the Regional Consultation Bangkok, Thailand, 7–9 January 2009. Avialable at: http://www.searo.who.int/entity/primary_health_care/documents/sea_hsd_320.pdf. Retrieved: 6, 2016.

[6] L. Trevena, H Davey, A. Barratt, P. Butow, and P. Caldwell, "A systematic review on communicating with patients about evidence". Journal of Evaluation in Clinical Practice, 12,(1): 13-23, 2006.

[7] J. Puustjärvi and L. Puustjärvi. "Practicing Evidence-Based Medicine in Developing Countries". In the proc. of the IST-Africa 2015 Conference & Exhibition. 2015.

[8] N. Mulyar, M. van der Aalst, and M. Peleg, "A pattern-based analysis of clinical computer interpretable guideline modeling languages," Journal of American Medical Information Association, Vol. 14, no. 6, pp.781–787.2007.

[9] M. Peleg, S. Tu, P. Ciccarese, and J. Fox, "Comparing computer-interpretable guideline models," A case-study approach. Journal of American Medical Information Association, 2003, pp.10–58.

[10] T. Leaon, K. Kaiser, and S. Miksch, "Free and open source enabling technologies for patient-centric,guideline-based clinical decision support," A survey. Methods of Information in Medicine, pp. 74–86, 2007.

[11] Baeza-Yates, R. and Ribeiro-Neto, B. 1999. "Modern Information Retrieval". Addison Wesley.

[12] "SPARQL Query Language for RDF", Available at: http://www.w3.org/TR/rdf-sparql-query/. Retrieved: 2, 2016.

[13] "RDF – Resource Description Language". Available at: http://www.w3.org/RDF/. Retrieved: 2, 2016.

[14] T.R. Gruber, "Toward principles for the design of ontologies used for knowledge sharing." Padua workshop on Formal Ontology, March 1993.

[15] G. Antoniou and F. Harmelen, "A semantic web primer," The MIT Press. 2004.

[16] M. Daconta, L. Obrst, and K. Smith, "The semantic web. :A Guide to the Future of XML, Web Services, and Knowledge Management," John Wiley & Sons. 2003

[17] "OWL - WEB OntologyLanguage", Available at: http://www.w3.org/TR/owl-features/ Retrieved: 2, 2016. Retrieved: 2, 2016.

[18] DuCharme. Learning SPARQL. O'Reilly Media.

Roadmap for Collaborative eHealth Service Architectures for Homecare

General eBusiness Requirements

Assoc. prof. Niels F. Garmann-Johnsen, Ph.D
Dept. of Information Systems,
Faculty of Social Science, University of Agder
Kristiansand, Norway
e-mail: niels.f.garmann-johnsen@uia.no

Assoc. prof. Santiago Martinez, Ph.D
Dept. Nursing and Health Sciences
Faculty of Health and Sport Sciences, University of Agder
Grimstad, Norway
e-mail: santiago.martinez@uia.no

***Abstract*— In many industrialized countries, such as the Scandinavian, one of the main goals for eHealth innovation is to use information and communication technologies to avoid the approaching inflection point in healthcare: it is estimated that by 2025 the demand of care could not be sufficiently attended based on today's healthcare delivery. eHealth services and business models for enabling older people in need of care and patients with chronic conditions may increase efficiency of healthcare and improve Quality of Life for residents. However, eHealth business models are known by their fast speed of change, what results in the need of a methodology to rapidly adapt technical requirements to these typical inter-agency changes. This work presents a research roadmap for collaborative eHealth Service Architectures for homecare. The research focused on the business requirements in service architectures (eBusiness, Enterprise systems), whose findings can inform eHealth research in general, health professionals and vendors.**

Keywords-eHealth; Service architecture; Business models; Enterprise systems; technology innovation; technology adoption.

I. INTRODUCTION

In many industrialized countries, such as Norway, one of the main goals for eHealth innovation is the use of information and communication technologies (ICT) and information systems (IS) to avoid the inflection point in healthcare: it is estimated that by the year 2025, the demand of care could not be sufficiently covered based on today's healthcare delivery [1]. Change and innovation are needed for treatments maintaining the quality of care within assigned budgets, where the resources needed transcend those of one singular organization. Technology that helps people with care needs to stay safely, longer at home and out of institutions and hospitals may be a strategy to reduce the total healthcare costs of society [2].

The authors present a roadmap for service design and architecture development in collaboration across organizations in healthcare. The research focuses in the home care, leaving clinical care outside the scope of this study. A challenge associated to the home care is that it presents a broad spectrum of user requirements within the same information technology (IT) infrastructure. Research in eBusiness models may inform future strategies and research in eHealth service design, especially in homecare, see Fig. 1. In this first paper, the authors reviewed the literature related to (e-)business models and eHealth to then offer propositions on service design.

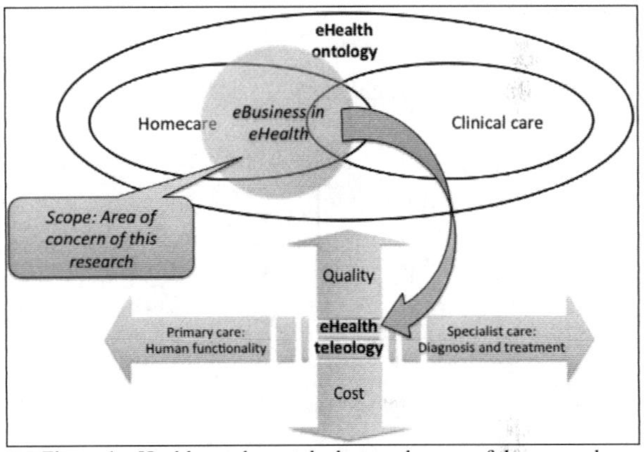

Figure 1. eHealth ontology, teleology and scope of the research

Studies of stakeholder requirements in eHealth innovation and pre-procurement [3] underline the need for increasing stakeholder involvement. Furthermore, design principles for business models that can facilitate such involvement are here elicited. The need in homecare is to develop an adaptive framework that allows front-line personnel (e.g., nurse, general practitioner) and other end-users to dynamically manage and (re-) configure business processes, patient services and health treatment clinical pathways, with the purpose creating service-oriented solutions whose core elements are patient autonomy and interactivity.

The authors have designed a three-step research roadmap with a peer-reviewed research article associated to each step (see Fig. 2). This article represents the first step, where we present the literature review on (e-) business models in the domain described followed by general business propositions. In the second step, a case study on the process of eHealth innovations will serve as a test of the elicited general (e-) business propositions and requirements of a design-process through action research in projects in a

Living Lab research methodology. In the third step, we seek to generalize the findings from the second step into a unified architecture or design model: an innovation and pre-procurement roadmap towards a technical architecture for a dynamic service management infrastructure that meets these requirements, enabling co-creation of eHealth services with end-users. Our level of analysis in the organization varies with each step. In the first one, we take a holistic view of the business model. In the second one, we focus on transitions in end-users' life, downstream activities and their consequences for collaboration with eHealth technology-vendors. In the third one, we apply findings from the two previous steps to induce a general design model for the architecture of ICT applications and infrastructure integrations, see Fig. 3 at the end of this article.

The main research question in this article is to find in the research literature general business requirements that influence eHealth services in home care. This lead to the two research questions addressed in this paper:

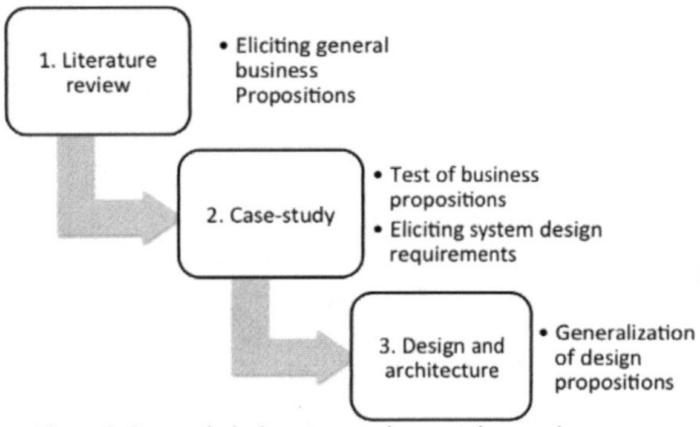

Figure 2. Research design, steps and expected research outcomes

RQ 1. What are the business requirements for Enterprise Systems that create value for users in the context of eHealth?

RQ 2. What are user preconditions to gain benefits from new Enterprise Systems and increase efficiency and quality in eHealth adoption?

The paper is structured in different sections. A theoretical framework is presented in Section 2. The method employed is described in Section 3. Section 4 presents the results, with the answers to the research questions. The paper ends with the conclusion and future work in Section 5.

II. Theoretical Framework

There is a variety of business models, but Hwang proposes three main types or business model templates that explain the business processes of modern enterprises better than the traditional Value Chain model of the firm [4] does alone [5]:

1. Solution shop. A place where customers describe their problem and it is fixed. For example, general practitioners and hospitals curing patients.
2. Value-adding process (comparable to the Value Chain) typical of manufacturing enterprises; comparable to Porters' [4] model of the firm. For example, pharmaceutical producers.
3. Facilitated user networks (Value Networks) to connect users and services. For example, different web communities (most communication- and internet-based businesses fall under this category).

With regard to these models, Hwang [5], building on works by Stabell and Fjeldstad [6], states that the facilitated user network is under-represented in the health sector—and, thus, represents growth potential: as more businesses move online, facilitated user networks represent a business model template of increasing significance, including eHealth. Thus the authors will explore these concepts further in relation to eHealth service architecture. Frameworks for analyzing value creation in e-Business (e.g., Amit and Zott, [7]) are relevant for understanding the eHealth landscape because many eHealth systems contain elements of e-Business models which belong to the category of Value Networks.

Amit and Zott's [7] e-business value creation model categorizes the sources of value creation into four main groups: *efficiency*, *novelty*, *lock-in* and *complementarity*. These groups are relevant to e-business models that often implement instances of the facilitated user network archetype of business models, and inter-organizational process collaborations. According to Amit and Zott, *efficiency* has been one of the primary driving forces for the transition to a more digital economy in all phases of the transaction. Complementarities that arrive from the bundle of goods provide more value to the users than the sum of values provided by each good alone. *Complementarity* and novelty imply collaboration and innovation—both themes of significance in healthcare. *Novelty* means that there is a "first" here, either in the service itself or in the way that it is conceived or delivered. Combining or connecting two or more previously unconnected parties, domains or methods often achieves this. A *lock-in* value is created through users' reuse of one-time transactions costs (e.g., for registration and service customization).

Seddon et al [8] proposed a model that explains how organizations, such as municipalities, can overcome organizational inertia and successfully adopt new enterprise systems and obtain benefits, e.g., quality and efficiency.

Enterprise systems support processes, information flows, reporting and business analytics within and between organizations and individuals. Specifically, these processes support organizations' business models. Business models are models of how enterprises create value for themselves and other, while also capturing value. An organization can have more than one business model, and business models can include more than one organization.

Since RQ 2 aims to explore the affordance of ES in realizing emergent business models in eHealth, the authors have chosen to use Seddon et al.'s model of key factors affecting Organizational Benefits from ES (OBES) [8]:

1. Short term/On-going business improvement projects:
    a. Functional fit (FF)
    b. Overcoming organizational inertia (OOI)
2. With the addition of long-term organizational benefit factors:
    a. Integration
    b. Process optimization
    c. Improved access to information

Functional fit of software means that the processes supported by the ES are efficient and effective for the organization and that they help people to finish their tasks. Overcoming organizational inertia is the extent to which organizations' members are motivated to learn, use and accept new systems. Integration refers to the unification of systems and/or data from resources within and outside an organization. Tailoring the information systems to an organization's goals help to achieve process optimization. This often necessitates improved access to information.

III. METHOD

The authors searched through scientific literature about business models in eHealth using Amit and Zott's framework [7] for value creation in eBusiness to answer RQ 1., and conceptualize the requirements. Then, we analyze concepts found in the review with the model proposed by Seddon et al [8] to answer RQ 2., and produce general propositions for information system design with special regards to primary care and homecare.

The literature search was performed in spring 2014 using the search engines ProQuest and EBSCOhost. The search engines targeted science and social science disciplines. The authors used truncations *health* AND (*ecommerce* OR *reward*) AND "business models" in the search terms. The search included scholarly papers and trade journals, choosing to include the latter due to the "newness" of the area in explored. The authors found 286 papers that matched the search criteria. However, a screening showed that 246 of the found papers were not related with eHealth business models. These papers were excluded. The excluded papers were concerned with topics such as: clinical trials, environmental issues, other non-health-related issues and management issues at large.

The remaining 40 papers (34 found through ProQuest and 6 found through EBSCOhost) were rated as relevant to the research topic and, thus, analyzed. A table showing the results and list of the searched and reviewed articles is included in the appendix.

IV. RESULTS

A. Answer to RQ1

This section shows the findings from the systematic literature review (for details, see appendix), which were aggregated and conceptualized following the guidance of Webster and Watson [9], and then sorted according to Amit and Zott's model. The authors added the category "Other" to collect concepts that were not easily sorted under Amit and Zott's four main sources; these were predominantly societal and public policy factors. These other sources stemmed from the special public or semi-public financial context of eHealth in countries like the British and Scandinavian.

1) Efficiency concepts

Easier scaling and sharing: new eHealth concepts leverage the efficiency and cost-flexibility of cloud computing. Cloud computing facilitates the use of intelligent algorithms to detect heightened risks of worsening conditions and supports the triage classification of patients (triage is an assessment of urgency of treatment). Such capabilities can be shared and utilized by several organizations. Standardizations and reductions in the number of product lines greatly reduce overhead cost ratios. Companies must defend low costs against overly personalized service.

Transaction cost efficiency: through e-business' transaction efficiency, doctors can now charge costs associated to phone calls and online consultation. Cloud computing also creates the opportunity for self-service solutions. Online solutions also enable easier care and quality coordination.

Easier sourcing: cloud services provide easier possibilities for sourcing support functions (e.g., system running).

2) Novel concepts

Rapid customization and co-innovation: the cloud model allows for rapid prototyping and innovation in processes, products and services. Co-innovation and co-creation generate new organizational and shared values. Surplus innovation can be sold-off, generating rent (commercial profit). Moreover, the possibility of doctor "visits" via secure video-conferencing enthuses patients. Partnerships with educational providers provide focused and personalized professional training.

3) Lock-In concepts

Revenue and reward splits: cloud services can store user preferences, which can enable product and service customization. Offering consumers, a choice of market channels can be mutually beneficial for all channels, due to the aggregation of products and services. For instance, the Hello Health website [10] negotiates a revenue split of subscriptions with physicians. Sharing information across organizational borders creates better possibilities for personalized health coaching.

Fun and belonging: online solutions can leverage the network and "gamification" (i.e., comparing progress with peers) effects of social media. Algorithms for the content and sentiment analysis of entries in online communities can suggest alternatives or interest groups that might be beneficial to join. Corporations should link medical plans for employees with corporate strategies, which could, for example, reward the use of fitness facilities and programs.

4) Complementarity concepts

Aggregating services: Internet offers possibilities for aggregating and sourcing services and finance. For instance, a combination of many travel-related and medical businesses and services can facilitate medical tourism).

Shared development risk: alliance management capabilities have positive effects on research and development capabilities. For example, in targeted medicine, there is a call for cross-business collaboration. Targeted medicine makes it possible to treat rare, previously untreatable conditions; however, the cost per patient is high, and some governments may refuse reimbursement. By reducing development time through improved collaborations between test laboratories and pharmaceutical companies, targeted medicine can be achieved at manageable costs. Pharmaceutical companies must learn how to fit into other companies' business models. All actors need to be satisfied, for example with "win-win" scenarios.

Management coordination: primary and secondary care must be better coordinated. Companies' external and internal resources must be aligned in order to facilitate the internalization of knowledge. Many user networks' business models are without profit or low profit; however, as a potential source of empowerment of patients, they represent ideal opportunities for business models for treating chronic diseases. These are costly treatments and are not well suited to the traditional business model of hospitals and physicians.

5) Other effects: Public and Societal concepts

New reimbursement schemes for better coordination: payment models in health should encourage coordination. Reimbursement models should encourage preventive measures, such as Healthy Outlook [11]. They should also reward the results of coherent treatment over "pay-per-consultation" systems. New reimbursement schemes may pose new challenges for financial reporting.

Knowledge management: knowledge benefits from exchange, rather than protection. The best "players" build "invisible" information infrastructures for developing, for example, cancer-focused IT system and evidence-based medicines.

Ethical compliance: Ethical compliance management is of growing significance in a changing eHealth business environment. Better management can be achieved through collaboration and democratic governance.

B. *Answer to RQ2*

Based on the business concepts induced from literature and rendered in the previous section, the authors propose the following propositions as design principles for a Municipal eHealth architecture:

Proposition 1. Shared e-commerce solutions: E-commerce solutions shared among collaborating partners, such as Hello Health [10], enable easier scaling and sharing, greater transaction cost efficiency and easier sourcing of non-strategic support functions, such as system running (i.e., application-server management). Shared e-commerce solutions support short-term functional fit and long-term business benefits, providing integration, process optimization and improved access to information.

Proposition 2. Co-creation with partners and clients (and patients): by providing customers and partners alternatives for customizing their services and by encouraging and proactively collecting feedback on services, it is possible to distinguish patterns that provide an impetus for improving services and creating new ones. Co-creation may produce positive short-term effects and long-term effects.

Proposition 3. Aligning internal and external resources: to overcome organizational inertia, structural changes may become necessary. Internal organizational resources should be dedicated to match external resources (business partners) in order to extract and embed external knowledge within an organization. This process is linked to propositions 4, 5 and 7.

Proposition 4. Aligning price and reward (bonuses and revenue sharing): aligning rewards and bonus schemes through the delivery chain and throughout user levels encourages the achievement of goals for treatment and health policies.

Proposition 5. Iterations for the economic alignment of interests: throughout a delivery chain, there is information asymmetry with regard to knowledge of processes and costs. Over time, all actors in the chain experience a learning curve; thus, collaborative iterations are needed to optimize business processes and to maintain win-win situations throughout the delivery chain. The role of ES is to provide accurate, reliable and comparable data for collaborative decision-making, thereby stimulating and supporting organizational learning.

Proposition 6. Social media integration and capabilities: to ensure that users and patients choose digital solutions and achieve the efficiency benefits that such solutions provide, the digital solutions should ideally be socially oriented and easy and fun to use. Gamification allows users to compare their personal data in fitness or treatment programs—and, thus, encourages and motivates users/patients to adapt and use solutions. Gamification can be developed and stimulated through social networks. Actors in healthcare (e.g., hospitals, primary care) need to develop their social media strategies [12].

Proposition 7. Alignment of incentives: a strong reallocation of Intellectual Property Rights and revenue sources may be necessary to encourage more efficient research and development and to achieve joint goals for development, thereby lowering the costs of new treatments (e.g., targeted medicine). Increased use of public-private collaboration contracts around research and development might be potentially considered.

Proposition 8. Transparency (sharing information across different levels): often, the patient is not the one who pays the full price for a medical service. In many economies, the state or voluntary or compulsory insurance or company medical plans pay for such services. Demographic changes, combined with improving treatment options, have led to rising costs of healthcare. To counter this, reimbursement schemes increasingly seek possibilities to pay for results rather than consultations. For health providers, this calls for increased information sharing, and transparency of treatment results. This proposition is linked to proposition 9.

Proposition 9. Linking performance measurement and finance: the enterprise/planning systems of medical providers and vendors must increasingly be able to provide data for quality performance and financial performance to both the market and the government. Business intelligence and balanced scorecard systems are needed for quality and performance management, and they can be directly linked to reimbursement schemes in the future.

Proposition 10. Democratic governance: ethical compliance is gaining importance in many areas, and it has great significance in eHealth [13]. A proactive strategy for compliance management, involving consistent policies for handling legal, quality and treatment deviations or complaints, including privacy and security issues, may benefit an organization in the long term. This may help overcoming barriers to e-Health adoption [14]. Actively involving users and employees in this work through democratic governance can enable such a result [13].

V. CONCLUSION AND FUTURE WORK

This study has answered the two research questions (RQs) by eliciting business models and enterprise systems design principles that may inform future eHealth architectures and applications. Although these principles are not exhaustive, they illustrate the need for flexibility and capability of co-creation with users at all levels.

In the future, the authors will test the found propositions in a Living Lab test setting [15], with special regard to adaptability, flexibility and facility for user-co-creation in design of eHealth services. Based on the empirical findings from these case-studies, we will seek to generalize a pre-procurement design process-model; a roadmap, in step three, which allows for a co-creation process that combines the goals of all stakeholders; application-vendors, municipalities and their constituents and end-users (see Fig. 3). The connection between innovations in local media and applications and central information infrastructures

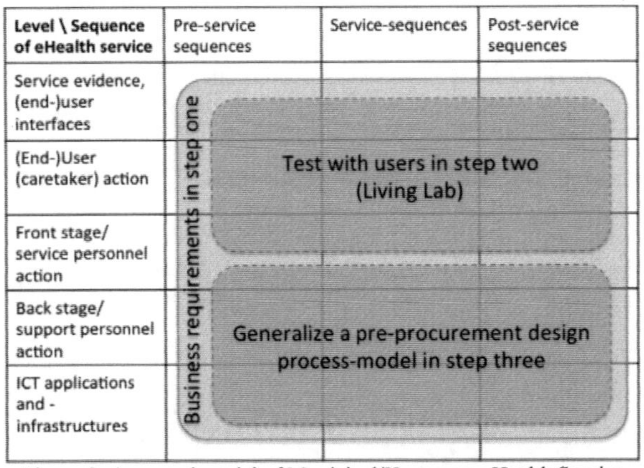

Figure 3. A general model of Municipal/Homecare eHealth Service Design: Level of analysis in each research step

(Software as a Service, cloud computing solutions) will also be explored.

REFERENCES

[1] D. M. Gaba, "The future vision of simulation in health care, " Simulation in Healthcare 2, no. 2, pp. 126-135, 2007.

[2] N. F. Garmann-Johnsen, "What seems to be the problem?—a study of connections between national contexts and regional e-health strategies," Health Policy and Technology, vol. 4, issue 2, pp. 144-155, 2015.

[3] N. F. Garmann-Johnsen, Ø. Hellang, editors, "Collaborative Process Modelling and evaluation in e-health," Scandinavian Health Informatics, Norway, 2014.

[4] M. E. Porter, "Towards a dynamic theory of strategy," Strategic management journal, vol. 12, suppl. 2, pp. 95-117, 1991.

[5] J. Hwang, "The innovator's prescription: an examination of the future of health care through the lenses of disruptive innovation," Archives of pathology & laboratory medicine, vol. 133, issue 4, pp. 513-20, 2009.

[6] C. B. Stabell, and Ø. D. Fjeldstad, "Configuring value for competitive advantage: on chains, shops, and networks," Strategic management journal, vol. 19, issue 5, pp. 413-437, 1998.

[7] R. Amit, and C. Zott, "Value creation in e-business," Strategic Management Journal, vol. 22, issue 6-7, pp. 493-520, 2001.

[8] P. B. Seddon, C. Calvert, S. Yang, "A multi-project model of key factors affecting organizational benefits from enterprise systems," MIS quarterly, vol. 34, issue 2, pp. 305-328, 2010.

[9] J. Webster, and R. T. Watson, "Analyzing the past to prepare for the future: writing a literature review, " MIS quarterly, pp. xiii-xxiii, 2002.

[10] R. L. Kravitz, "Improvement happens: a commercial IT solution for reviving primary care? An interview with Hello Health CEO Nathanial Findlay and colleagues," Journal of general internal medicine, pp. 1-5, 2013.

[11] P. Micheli, M. Schoeman, D. Baxter, K. Goffin, "New business models for public-sector innovation: successful technological innovation for government," Research-Technology Management, vol. 55, issue 5, pp. 51-57, 2012.

[12] R. Effing, and T. A. Spil, "The social strategy cone: towards a framework for evaluating social media strategies," International journal of information management, vol. 36, issue 1, pp. 1-8, 2016.

[13] C. L. Rossi, "Compliance: an over-looked business strategy," International Journal of Social Economics, vol. 37, issue 10, pp. 816-831, 2010.

[14] I. Pirillo, G. Ricci, F. Amenta, "The Future of e-Health," The Sixth International Conference on eHealth, Telemedicine, and Social Medicine, eTELEMED 2014, pp. 280-284, 2014.

[15] M. Pallot, B. Trousse, B. Senach, D. Scapin, editors, "Living lab research landscape: from user centred design and user experience towards user cocreation," First European Summer School Living Labs, 2010.

A Qualitative Study of Parents` Experiences of Information Exchange between the Tertiary Health Service and the Child`s Local Support Team using Videoconference

Merete Kristin Tschamper (MPhil)
Dept. of children and youth, Division for Neuroscience,
National Center for Epilepsy
Oslo University Hospital
Oslo, Norway
E-mail: meretek@ous-hf.no

Wenche Schrøder Bjorbækmo (PhD)
Dep. of Health Science
The University of Oslo
Oslo, Norway
E-mail: w.s.bjorbakmo@medisin.uio.no

***Abstract* - In this study, four children with rare, epileptic condition including intellectual disability were interdisciplinary assessed in the National Center for Epilepsy (SSE). The results of the assessments were communicated to the children`s local support service using videoconference (VC) with the parents present. *The purpose* for this study was to obtain a deeper insight into parents` experiences with information exchange using VC. *Methods:* In-depth interviews were performed with five parents within a week after the VC. The generated material was analyzed with a thematic hermeneutic research approach. *Results:* The local participation increased from 2-6 in traditional face-to-face meetings to up to 22 participants in the VCs. The parents described how the increased participation ensured a direct, oral information exchange that prevented treatment misunderstandings. The parents were relieved from the responsibility to inform all professionals involved in the local community. In addition the parents suggested this increased the multidisciplinary collaboration and facilitated a comprehensive approach to the children`s complex situations. The parents experienced increased user-involvement in the process of arranging the VC compared to traditional face-to-face consultants. Some parents had started to initiate VC themselves. *Conclusion:* VC was considered a more potent tool for information exchange between providers than face-to-face meetings due to the increased local participation and that everyone was assured to get the same information simultaneously. However a crucial prerequisite seem to be that the professionals from the tertiary health service had clinical knowledge of the child and tailored the information exchanged according to each child and family`s needs.**

Keywords-Parent; chronic disease/illness; epilesy; disability; videoconference; patient education.

I. INTRODUCTION

Epilepsy in combination with disability is a condition that causes various needs for long-term multidisciplinary support. Research show that parents with chronic ill children experiences the lack of information exchange between the child`s providers as one of the main stressors leading to inadequate treatment of the child [1]-[3]. Among other services SSE provides a nationwide and interdisciplinary assessment- and treatment service for people with rare epileptic conditions. A disease is defined rare in Europe when it affects less than1 in 2000 of the general population [4]. Each condition affects the patient individually and there is a need for tailored treatment. SSE, together with The National Center for rare, epilepsy related disorders (NK-SE), are legally required to educate the patients, their relatives, and the patient`s local support service in order to ensure optimal treatment [5][6]. Norway is aerially a relatively large country with a widely distributed population. Communication between patients and providers using VC bridge geographical distances and thus holds the potential to be cost-efficient with a broader multidisciplinary scope for information exchange than traditional face-to-face consultants [7]-[10]. Although SSE and NK-SE increasingly use VC to exchange information as an alternative to traditional face-to-face meetings, scant attention has been paid to the parents` experiences with VC as an arena to exchange information about their child`s condition and care [11]-[13]. There is a need for a deeper understanding of how parents of chronic ill children experiences information exchange using VC. A deeper insight into the parents` experiences may have implications for adjusting standard practice to better meet their needs. The findings are likely to be transferable beyond the current context.

II. INTERDISCIPLINARY ASSESSMENT

An interdisciplinary assessment in SSE normally contains 17 days of hospitalization. According to each child`s individual need, this might include a medical assessment performed by a pediatrician and a nurse specialist in epilepsy, a cognitive assessment performed by a psychologist and a special educator, a motor assessment performed by a physiotherapist and an occupational therapist. The parents also have the opportunity to consult a social worker while hospitalized for advice concerning legal rights for benefits and support. At the end of the hospitalization the results of the assessment, along with recommendations for treatment and educational interventions are communicated to the parents and the child`s local support service either in a meeting face-to-face at SSE or using VC.

III. VIDEOCONFERENCE

VC is a virtual way of meeting that can be used to exchange information, and provide support and care. The security of sensitive information is ensured through the Norwegian Health Net [14]. In this study the local participants met in VC rooms at the municipal Labour- and Welfare departments. The professionals from SSE or NK-SE participated from a VC-room at Oslo university Hospital. The VCs lasted from 1, 5 to 2, 5 hours.

A. *Participants in the VCs*

The parents participated in the VC concerning their child. They had the choice whether to participate while hospitalized at SSE or at home after discharge. The multidisciplinary professionals responsible for assessing each child participated from a VC-room at SSE.

The special educators responsible for the educational assessment were responsible for arranging the VC in cooperation with the parents. The parents decided whom to invite locally. The local participants were multidisciplinary supporters from the children's local support service. According to each child`s individual need, that might include teachers and assistants, municipal physiotherapists, health visitors, representatives from the municipal educational and psychological counseling service, social educators and assistants from respite care-homes. In one VC a grown up sibling, an aunt together with representatives from the municipal Labour-and Welfare department participated.

B. *Content of the VCs*

The results from the interdisciplinary assessments, together with treatment options, were presented and discussed with the local participants in the VC. Another goal for the information exchange was to discuss possible facilitation of the child`s environment to prevent seizures, contribute to optimal development and quality of life. The parents were present in the VC. According to practice they were informed about the results of the assessments in advance of the VC.

IV. METHODS

The empirical material stems from in-depth interviews exploring the parents' experiences of the VC. The thematic hermeneutic analyses of the interview identified several themes of which three is presented in this article.

A. *Inclusion criterions*

The parents participating in the study have a child aged 5-12 with a rare epileptic condition that include moderate to severe intellectual disability. The study included 5 parents who`s child had recently undergone an interdisciplinary assessment at SSE followed by an information exchange with the local support service using VC. In order to provide a nuanced description of specific cases, the parents fitting the inclusion criterions were strategically chosen consecutively as they were admitted to SSE. Only one parent had experienced information exchange using VC at SSE before. Three of the parents had experienced multiple traditional face-to-face meetings in SSE exchanging information with the child´s local support service. One of the children had been hospitalized in SSE for the first time.

B. *Collecting of data and analyses*

In-depth interviews were performed with four mothers and a father within a week after the VC. The interviewer did not attend the VCs. The interview guide had few and open questions that aimed to explore how the parents experienced the use of VC for information exchange. The interviews were transcribed verbatim and analyzed with a thematic hermeneutic research approach [15]. Three of the initial themes identified were: A) VC an efficient tool to exchange information between multidisciplinary providers B) Increased parental control and involvement C) The impact of the relationship with the professional arranging the VC.

V. FINDINGS

In the following, the results from three of the identified themes will be presented.

A. *VC an efficient tool to exchange information between multidisciplinary providers*

In the VC there were from 10 to 22 local participants compared to 2-6 in former face-to-face meetings at SSE. According to the parents, the direct information exchange and discussions between the professionals prevented treatment misunderstandings that occurred when the local support service only received written reports. In addition, the parents experienced an increased multidisciplinary collaboration that facilitated a comprehensive approach to the child`s complex situation. Finally the parents saved time not having to attain meetings explaining the reports from SSE to the local providers.

B. *Increased parental control and involvement*

The parents experienced increased control and user-involvement in the process of arranging the VC. As almost all the invited local professionals participated, the scope for the information exchange widened. Some of the parents actively used the VC as an arena to inform not only the providers of care and education, but also other providers like social workers, family members or the principal on the local school. Some parents had started to initiate VC themselves by contacting personnel at SSE or NK-SE.

C. *The impact of the relationship with the professional arranging the VC*

The parents suggested that the positive and trusting relationship they had established with the professional arranging the VC was important for the quality of the information exchange. They also considered it crucial that the professional from SSE and NK-SE had clinical knowledge of their child in addition to special scientific knowledge of the rare, epileptic conditions in order to tailor the information to the child and family`s needs.

VI. Discussion

For the parents in this study the most important issue was that the VC increases the local participation. The increased local participation enabled a direct information exchange between providers from different levels of the health service, school and social services in a way that would not have been possible without this technology. Thus The VC offers an arena for the participants to relate, share experiences and co-shape their knowledge of the child´s situation and needs. The possibility for the local providers to ask clarifying questions directly to the professionals prevented misunderstandings in the follow-up treatment.

However, this outcome seemed to depend upon the quality of the interaction created between the participants in each VC. As the interactions in the VC are dynamic, the parents experienced the quality of the interaction differently from each other. To ensure that misunderstandings in the follow up treatment are prevented, the participants need to contribute and ask question – this requires a dialogue. Without dialogue the VC becomes an arena simply for information transfer – not for information exchange.

Whereas some of the VCs in this study seem to have a relaxed ambiance where the local participants asked questions and contributed, others seem to have been tense, with less interaction. The parents suggest the local participants might feel insecure talking in the virtual space in front of a camera with the whole local support team present. The VCs with the most dialogue seemed to be the ones were the participant communicated at regular basis using VC. The communication in the VC where the participant met for the first time a mother described as "a monologue." In this VC the local participants listened without responding. If the local providers do not provide any feedback, it is impossible for the participants from SSE or NK-SE to correct potentially incorrect impressions. Thus, it is possible that the VC might create, not prevent, misunderstandings.

The parents' relationship with the providers seemed important for their experiences of the quality of the information exchange. It seemed crucial for the parents to have a trusting relationship with the professional arranging the VC, and for them to share the same opinion regarding the child`s condition. According to the parents, the professionals organizing the VC recognized and used the parents' experiential knowledge of the child as a basis for the information exchange. The parents' opinion and the professionals' clinical understanding of the child´s situation seem to have united during the hospitalization in SSE/SS.

The parents explained that they trusted the professionals at SSE or NK-SE to describe the child`s needs in a tailored way. They also appreciated the way they were included in all stages of organizing the VC. This, and the fact that they decided whom to invite locally, seemed to increase the parents' sense of control of the information being exchanged in the VC.

The parents also appreciated that the child was interdisciplinary assessed during one hospitalization. This facilitated a comprehensive approach to the child and family`s needs.

VII. Conclusion and Further Work

The parents considered the VC a more potent tool for information exchange between providers than face-to-face meetings due to the increased local participation. However a crucial prerequisite seem to be that the professionals from SSE and NK-SE had clinical knowledge of the child and tailored the information exchange according to the child and family`s needs. In this qualitative study the sample was not designed to be statistically representative. However the findings are consistent with other research concerning the use of VC in information exchange for patient with long-term conditions [7]-[10], [16]. The importance of a trusting relationship between the patients and the providers for the interaction to be successful is also consistent with previous research [17][18]. There is a need for further studies of the conditions that needs to be present in order to ensure the quality of the information exchange using VC. The long-term effect on the treatment of the child and whether the use of VC reduces hospitalizations also needs to be investigated.

Acknowledgment

We would like to thank the parents and the providers who participated in this study. The study was conducted in collaboration with The National center for rare, epilepsy related disorders, Oslo University Hospital and Solberg School, Norway.

References

[1] Gundersen T. 'One wants to know what a chromosome is': the internet as a coping resource when adjusting to life parenting a child with a rare genetic disorder. Sociology of health & illness. 2011;33(1):81

[2] Grut L, and Kvam M. Facing ignorance: people with rare disorders and their experiences with public health and welfare services. Scandinavian Journal of Disability Research. 2013;15(1):20-32

[3] Pelentsov LJ, Fielder AL, Laws TA, and Esterman AJ. The supportive care needs of parents with a child with a rare disease: results of an online survey.(Report). BMC Family Practice. 2016;17(1)

[4] What is a rare disease? [Available from: http://www.rarediseaseday.org/article/what-is-a-rare-disease

[5] Pasient-brukerrettighetsloven. Pasient- og brukerrettighetsloven. LOV-1999-07-02-63: Helse- og omsorgsdepartementet; 1999 [Sist endret: LOV-2014-06-20-41 fra 01.01.2015:[Available from: https://lovdata.no/dokument/NL/lov/1999-07-02-63

[6] Spesialisthelsetjenesteloven. Spesialisthelsetjenesteloven. LOV-1999-07-02-61 Oslo: Helse- og omsorgsdepartementet; 1999 [Available from: https://lovdata.no/dokument/NL/lov/1999-07-02-61

[7] Cameron MPL, Ray R, and Sabesan S. Physicians' perceptions of clinical supervision and educational support via videoconference: a systematic review. Journal of telemedicine and Telecare. 2014, Vol. 20(5) 272–281

[8] Lindberg B, Nilsson C, Zotterman D, Soderberg S, and Skar L. Using information communication technology in home care for communication between patients, family members, and healthcare professionals: a systematic review. International journal of telemedicine and applications. 2013;2013:31

[9] Hagen K. Users' Experiences with Clinical Video Conferencing at a Resource Centre for Rare Disorders [Master]: University of Tromsø, Norway; 2010

[10] Gentles SJ, Lokker C, and McKibbon KA. Health information technology to facilitate communication involving health care providers, caregivers, and pediatric patients: a scoping review. Journal of Medical Internet Research. 2010;12(2):e22

[11] Chi NC, and Demiris G. A systematic review of telehealth tools and interventions to support family caregivers. Journal of Telemedicine and Telecare. 2015;21(1):37-44

[12] McLean S, Protti D, and Sheikh A. Telehealthcare for long term conditions. BMJ: British Medical Journal. 2011;342(7793):374-8

[13] Åkesson K, Saveman BI, and Nilsson G. Health care consumers' experiences of information communication technology: A summary of litterature. International Journal of Medical Informatics. 2007;76(9):633-45

[14] Norsk Helsenett SF. [Available from: https://www.nhn.no/helsenettet/video/Sider/default.aspx

[15] Braun V, and Clark V. Successful qualitative research: a practical guide for beginners. LA: SAGE. 2013

[16] Steel K, Cox D, and Garry H.Therapeutic videoconferencing interventions for the treatment of long-term conditions. Journal of Telemedicine and Telecare 2011;17(3) 109-117

[17] Wade V, and Eliott J. The role of the champion in telehealth service development: a qualitative analysis. J Telemed Telecare 2012;18:490–492

[18] Carlisle K, and Warren R. A qualitative case study of telehealth for in-home monitoring to support the management of type 2 diabetes. Journal of Telemedicine and Telecare 2013;19(7) 372–375

Kansei Design Applied on Hospital Beds: Medical Design for Patient and Caregiver

Eiji Onchi

Kansei, Behavioral and Brain Sciences
Graduate School of Comprehensive Human Sciences
University of Tsukuba
Ibaraki, Japan
Email: onchi.eiji@gmail.com

SeungHee Lee

Faculty of Art and Design
Graduate School of Comprehensive Human Sciences
Univeristy of Tsukuba
Ibaraki, Japan
Email: seungheekansei@gmail.com

***Abstract*—The paper proposes new design directions, using a Kansei Design approach, of the mechatronic hospital bed made by the Pontifical University Catholic of Peru, and analyzes the Kansei aspects of the overall user experience. Implementing a semi-autonomous, easy to use adjustable bed, not only facilitates the work of medical personnel, but may improve the recovery rate of patients by boosting their emotional state by enabling mobility. It is suggested to correlate the personality of the users with the impressions of this medical bed to find if using a mechatronic system is suitable for all patients.**

Keywords–hospital bed; kansei design; rehabilitation; user experience; emotions.

I. INTRODUCTION

The technology developed in the past years has eased the advancement of automation in different areas, meaning that the quality of life of humans could arguably improve. In spite of that, there are concerns that automation will dehumanize certain procedures or could be misused and abused if not used correctly [1]. In response to this, researches and developers are starting to apply Kansei Design approaches to consider human factor, with a focus on human emotions, as the core when designing a product. The use of 'kansei' (a Japanese term that conveys the meaning of affection, emotion, aesthetics, and related terms) as the base for developing a product, aims to improve the synergy between humans and objects [2].

In health-care, automation aims to be beneficial for both patient and caregiver. In this regard, the evolution of adjustable beds, commonly known as hospital beds, plays a key role in any medical institution. From the traditional mechanical and electromechanical beds (controlled by handles or buttons that are cumbersome to operate), to the most advanced automated medical beds (embedded with sensors, actuators and a central processing unit), the goal has been to provide movement and comfort to bedridden patients.

Nevertheless, it is of the uttermost importance to take into account injuries caused by prolonged lack of movement, such as skin pressure ulcers, and reduce the risk of death by sleep disorders, like sleep apnea, while relieving caregivers of work-related stress. In this sense, the University of New Hampshire developed a medical bed [3] that monitors the patients' blood pressure and breath rates in real time to react to possible complications. Moreover, beds that monitor temperature and spatial location of the patient [4] can prevent skin injuries, while systems with pressure sensors [5] help for apnea sleep disorder. Following this trend, hospital beds as those produced by EPOSbed project [6] or Stryker [7] implement mechatronic systems to ease the workload on the caregiver and yield more freedom to the patient.

In addition, to complete and complement the aforementioned researches and projects, the Pontifical Catholic University of Peru developed a mechatronic SmartBed [8] with these considerations in mind. However, most of these developments mainly focus on the technical features and usability of adjustable beds rather than the user experience and 'kansei' of the patient, the caregiver, and those members related to the bedridden person.

Therefore, based on this [8] previous research, this paper discusses the developed mechatronic hospital bed from a Kansei Design perspective and proposes improvements on its design. The structure of the document is as follows. In Section I, the State of the Art development of hospital beds are introduced. In Section II, the general characteristics of the SmartBed [8] are presented. Section III details the experimental procedure and obtained results, next to the analysis and discussion in Section IV. Finally, in Section V, the conclusions of this research are laid out.

II. MECHATRONIC SMARTBED

The following section briefly describes the characteristics of the mechatronic SmartBed show in Figure 1.

1) Mechanical Specifications: This hospital bed can be adjusted to seven standard medical positions, namely: trendelenburg, anti-trendelenburg, fowler, gatch, self-contour, lateral,

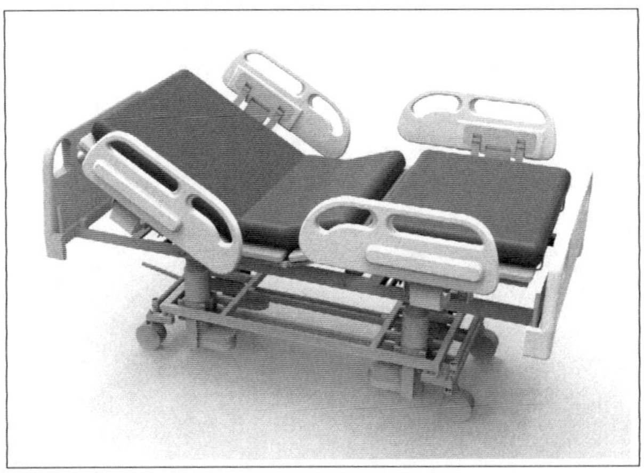

Figure 1. Mechatronic SmartBed.

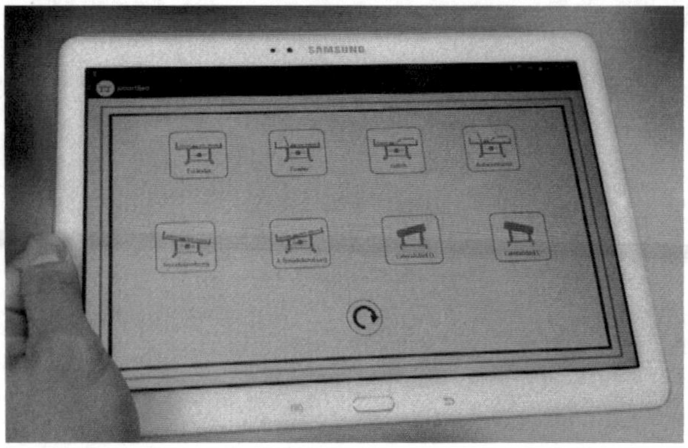

Figure 2. Smartbed User Interface

and flat. To achieve these shapes, four linear actuators act as pillars, while two linear actuators control the angle of the header and footer.

2) Electronic Specifications: Several sensors measure the weight, position and shape of the patient, while synchronizing these data to a web-based database. The bed itself communicates wireless to a tablet, where doctors and caregivers are able to manipulate the medical bed at ease.

3) Interface Design: All collected data is synchronized and stored on a web server in order to be accessible anywhere with an Internet connection. In this manner, family members, as well as medical personnel, can keep track of the patient's condition in real time. The interface is presented on a tablet and uses icons to control all motor functions of the hospital bed, as shown in Figure 2, while providing processed data from sensors. The tablet can only control paired beds, so as to avoid misuse or hacking.

III. Experiment and Results

A. Experiment

As stated in [8], the carried out user study had two objectives: 1) confirm the functionality of the system, and 2) evaluate the users' perception of the interface. In this regard, the focus of this paper will be to discuss the later. To accomplish the objectives, 20 participants from 17 to 28 years old tested the adjustable bed, as shown in Figure 3, and filled a survey that encouraged the participants to think of the system as a robot.

The procedure for the experiment went as follows:

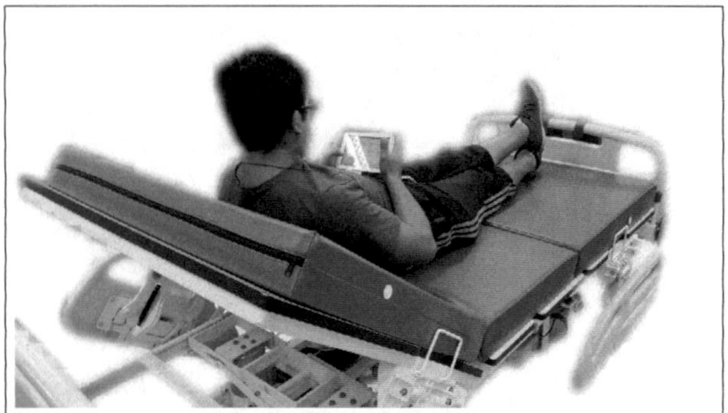

Figure 3. SmartBed during the experiment

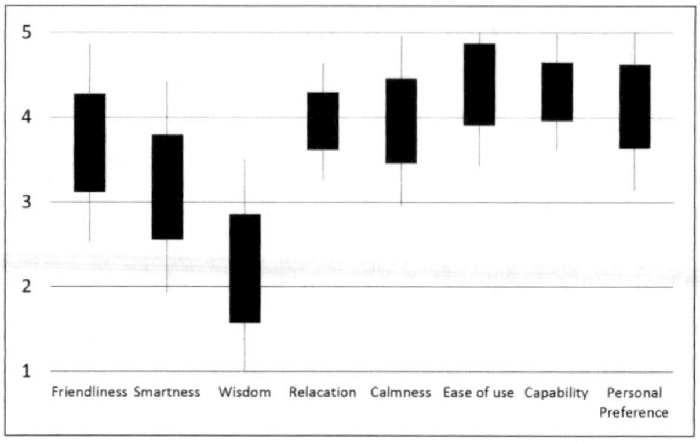

Figure 4. User Experience Survey

1) *Training (1 min)* - participants laid down on the bed and were given instruction on how to navigate the user interface.
2) *Predefined sequence (2 min)* - participants experienced automatic movement of the SmartBed.
3) *Free control (3 min)* - participants freely interacted with the interface.
4) *Questionnaire (4 min)* - participants filled out a questionnaire

For the survey, the following items were asked and answered with a scale from 1 (Not at all) to 5 (Of course), using a Godspeed style questionnaire [9]:

- *Friendliness*: how friendly the system feels.
- *Smartness*: how smart the system feels.
- *Wisdom*: how knowledgeable the system seems.
- *Relaxation*: how relaxed the system seems.
- *Calmness*: how calmed the system feels.
- *Ease of use*: ease of use of the interface.
- *Capability*: how capable the system feels for the task.
- *Personal Preference*: whether the subject would like to be catered by a robotic interface.

B. Results

The results of the survey presented in Figure 4 are as follows. Having an adjustable bed controlled by just a tablet seemed to be easy to use for the participants. Moreover, most subjects felt the bed to be calmed and relaxed. On the contrary, while some people thought the SmartBed was somewhat smart, the majority agreed that it did not infer wisdom.

C. Follow Up

A follow up of this experiment took place on an actual hospital environment, where a designated nurse and patient tested the bed for a week. From the care-giver's perspective, the developed system proved to be functional and convenient, as they could control the hospital bed remotely, without any major effort. On the other hand, the patient reported that the movement of the bed was comfortable and felt secure while using the system.

Notwithstanding, this convenience of controlling the SmartBed remotely might distance the interaction between caregiver and patient, while the emotional state of the bedridden may not be conveyed promptly and correctly.

IV. Kansei Analysis and Discussion

In Kansei Design, user experience plays an important role in defining the direction of development for a product. The reaction of the final user is as important as solving the problem with the most optimal procedure, which is the reason why a constant feedback from the users is necessary to create a closed-loop like system that allows for a better design. Additionally, the environment where the product will be used must be taken into consideration, as that determines limitations and restrictions of the design.

Taking the aforementioned factors into account, the results presented on the previous section show that the overall feedback of the participants tends toward perceiving the bed as having seemingly human-like characteristics. Nonetheless, it is worth noting that the survey encouraged the subjects to think of the SmartBed as a robot, instead of an entirely controllable machine. Moreover, the subjects were around their twenties, which means that it is highly probable that they are used to interacting with technology.

In this sense, the general perception that this hospital bed is capable of doing what it is expected to do, but without having any knowledge or good judgment for its actions, namely wisdom, could mean that potential users (both patients and medical personnel), might think of the system as a slave machine rather than a companion throughout rehabilitation. Yet, this perception may give the patient the notion of freedom and independence, as the user will be able to control the system by itself, thus potentially increasing his/her motivation and mood, and improve the rehabilitation process.

Furthermore, as the SmartBed is projected to be used by people not familiar with technology, it is proposed to redesign the user interface. With the bed's ability to do biometric measurements of the patient, it is possible to process these data to present some of the emotional states of the user, as well as programming countermeasures to react to negative emotions. Also, Figure 5 shows a draft of the redesigned user interface: a cover and tablet which displays the face of the patient and caregiver, alongside simple inputs, to lessen the negative effect produced by remote controlling, and to make the bed both easier to use for general people, and a communication's channel.

On the other hand, a redesign of this medical device, considering the approach suggested by Lee, et al. [2] of creating a 3D model in clay before sketching, can be applied to obtain unconventional results to actual hospital beds in order to circumvent the feeling of entrapment for bedridden patients. After that, the model must be adjusted with hospital safety standards to ensure its usability.

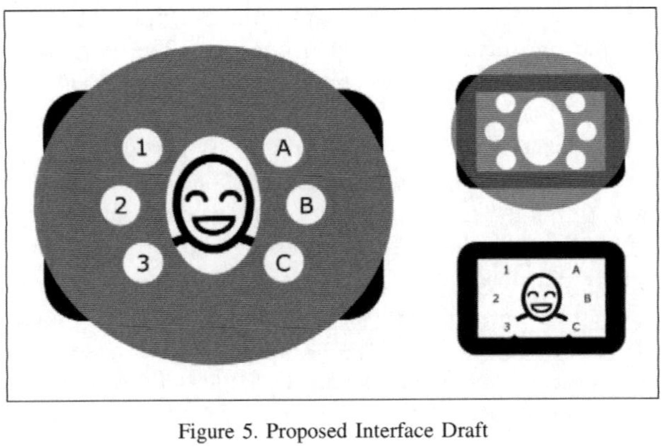

Figure 5. Proposed Interface Draft

V. Conclusion and Future Work

Although designing a hospital bed with only technical aspects in consideration may improve user experience, the 'kansei' of the user must be appraised to develop a medical device that feels and is safe.

Likewise, adding convenience to a system might improve usability, but could diminish patient-caregiver relationship. Thence, it is suggested to consider these factors on future designs.

Finally, further evaluations must be made on an actual hospital environment to assess the emotional and psychological response of target users. It is suggested to correlate the personality of the patient with the impressions of this SmartBed to find if using a mechatronic medical bed is suitable for all patients.

Acknowledgment

The authors would like to thank the Potifical Catholic University of Peru and its researchers for providing the background work of this project.

References

[1] R. Parasuraman, "Humans and Automation: Use, Misuse, Disuse, Abuse," Human Factors, vol. 39, no. 2, 1997, pp. 230–253.

[2] S. H. Lee, A. Harada, and P. J. Stappers, Design Based on Kansei. Taylor Francis, London, 2002, chapter 16, pp. 219–229, in Green, W. S. and Jordan, P. W., Pleasure With Products: Beyond Usability, ISBN: 0-415-23704-1.

[3] "UNH Research Enhancing Patient Care With "Smart" Hospital Beds," 2012, URL: http://www.unh.edu/news/cj_nr/2010/dec/dw07smart.cfm [accessed: 2017-02-26].

[4] A. Bandala and et al., "Development and Design of Automated Hospital Bed with Incremental Panels for Bedsore Prevention," in IEEE Region 10 Conference TENCON 2014, Bangkok, Thailand, Oct. 2014, pp. 762–767.

[5] D. Townsend, M. Holtzman, R. Goubran, M. Frize, and F. Knoefel, "Relative Thresholding With Under-Mattress Pressure Sensors to Detect Central Apnea," IEEE Transactions on Instrumentation and Measurement, vol. 60, no. 10, 2011, pp. 3281–3289.

[6] "Welcome to the EPOSbed Project Website," 2011, URL: http://www.eposbed.com/ [accessed: 2017-02-26].

[7] "In-Touch Critical Bed Care," 2016, URL: http://www.stryker.com/en-us/products/PatientHandlingEMSandEvacuationEquipment/Beds/Critical CareBeds/InTouch/index.htm [accessed: 2017-02-26].

[8] E. Onchi, C. Pealoza, and F. Cullar, "Design and implementation of a mechatronic SmartBed for improved rehabilitation," in Proceedings 2016 IEEE International Conference on Industrial Technology (ICIT), Taipei, Taiwan, Mar. 2016, pp. 1482–1487.

[9] C. Bartneck, E. Croft, and D. Kulic, "Measurement Instruments for the Anthropomorphism, Animacy, Likeability, Perceived Intelligence, and Perceived Safety of Robots," International Journal of Social Robotics, vol. 1, no. 1, 2009, pp. 71–81.

Sleep Evaluation Influenced by Negative Emotions and Memories After Awakening Among Adolescents and Young Adults

Huan Wen Lu
Faculty of Industrial Design
Eindhoven University of Technology
Noord-Brabant, The Netherlands
Email: klu@live.nl

SeungHee Lee
Faculty of Art and Design
Graduate School of Comprehensive Human Sciences
Univeristy of Tsukuba
Ibaraki, Japan
Email: seungheekansei@gmail.com

***Abstract*—This paper examined the influence of sleep experience on self-sleep evaluation among adolescents and young adults. 64 participants (age between 16-27, 36 males, 28 woman) have participated in a questionnaire study. Sleeping habits and subjective sleeping experience have been mapped and analyzed. In a follow-up study, three participants have been tracking their sleep experience, daily activities and sleep hygiene. The results have shown that the participants of both studies sleep less than average, but within the margin for healthy sleep. However, 25,8% of the participants evaluate their sleep as insufficient or as an unpleasant experience. The case study showed that the evaluation process of sleep experience is unbalanced. Memories of bad sleep and negative emotions have a bigger impact on sleep evaluation scores. Memories of pleasant sleeping experiences are more easily replaced by new thoughts and are often forgotten during the sleep evaluation process.**

Keywords*– *sleep hygiene; sleep evaluation; activity tracking; bad sleeping experience.

I. INTRODUCTION

Insufficient sleep, difficulties waking up and a bad mood are consequences of experiencing bad sleep. In the long run, unpleasant sleeping experience and mental fatigue may lead to sleeping disorder [1]. Out of this, only a small fraction of the big group of problem sleepers has officially been diagnosed with a sleep disorder and are receiving professional help and treatment [2]. The common sleeper is thus, unaware of what good sleep and bad sleep implies. On the other hand, information found on the internet is not always accurate or is difficult to understand for everyday people. As a result, the difference between good sleep and bad sleep is therefore often misunderstood [3]. Sleep can be influenced by many different factors, where sleep hygiene and consuming behavior plays an active role [4].

In recent studies regarding sleep quality perceptions, it has been suggested that there is a wide variety in how accurately insomnia sufferers perceive their sleep [5]-[8]. Insomnia sufferers tend to underestimate the overall sleep time and have a negative view on sleeping compared to normal sleepers. Furthermore, investigations into sleep misperceptions have shown that people with a healthy sleep hygiene have proven to be more accurate to evaluate their sleep [9]-[11]. In this regard, the goal of this study is to map down and analyze the sleep data of adolescents and young adults to find how self-sleep evaluation is influenced by sleep experience.

The topic of this paper is introduced in section I. In section II, the methodology, target audience and the setup of both studies is explained. Section III shows the obtained results and subject evaluation of both studies. The discussion and conclusion are written in section IV and V.

II. METHODS

This study is divided into two parts: (1) the sleep evaluation analysis of the younger generation in the area of Eindhoven and (2) a case study of three subjects to find how self-sleep evaluation is influenced by sleep experience. Adolescents and young adults were approached to participate in a survey and share their past experiences, thoughts about sleep quality and how they would evaluate their own sleep. Moreover, people regardless of background, gender, or medical dossier were allowed to participate in this questionnaire. The only requirement was to be within the age group of 16 to 27 years old.

According to the adolescents and young adults participating in this study, they are facing more problems related to sleep such as: feeling tired, insufficient sleep, irregular sleep patterns and stress. To look further into the case of sleep evaluation, a specific case study is designed to track the sleeping experience and consuming behavior of adolescents and young adults outside a clinical environment.

A total of 64 participants (28 female) successfully filled in the sleeping evaluation questionnaire. Figure 1 and 2 show the age distribution and occupation. The occupations are distributed among three groups: 66% are in education or training, 31% are employed, and 3% are unemployed and not in education. Among this group, 16 participants (25,8 %) stated they have troubles regarding one, or more of the following points: falling asleep, staying asleep, going to bed, waking up, or tiredness. In all cases the participants were not officially diagnosed with any sleeping disorder.

The following statements were hypothesized: (1) Adolescents and young adults do not have a clear overview of what good sleeping experience means. The true definition of good sleep is not clearly defined, causing misunderstandings; (2)

sleep evaluation is influenced by time. The moment when it is evaluated makes a difference in evaluation score. People have different opinions about their sleep during the day; (3) only a small percentage of the adolescents and young adults do have (symptoms of) sleeping problems; (4) consuming and activity behavior prior going to bed does have a measurable impact on sleep experience. Physical exercise, caffeine and alcohol consumption may have an influence on one's sleep.

Based on the results of the sleep evaluation analysis, three subjects were chosen out of the 64 participants to participate in a case study. They were chosen from the initial group of participants based on questionnaire results, Pittsburgh Sleep Quality Assessment (PSQI) score and willingness to participate. What is more, the subjects completed an interview process, about their personal standing on the subject of sleep, survey results and past sleeping experiences. They, consisted of two non-problem sleepers and one problem sleeper, where the last has not been diagnosed with a mental sleep disorder. In this research, the following assumptions were made: a non-problem sleeper is someone who (a) sleeps everyday at least six hours; (b) falls asleep in less than 30 minutes; (c) is sleeping in a bedroom environment; (d) is excluded from any disorders or illnesses; (e) has a PSQI score of less than five. A problem sleeper is defined if he or she does not qualify for at least one of the statements written above.

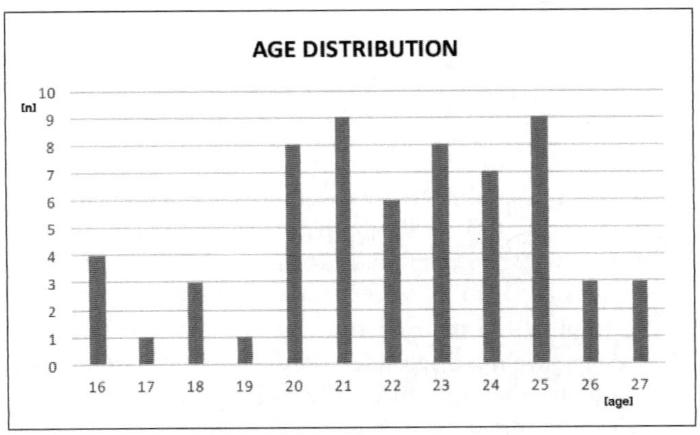

Figure 1. Age Distribution Participants

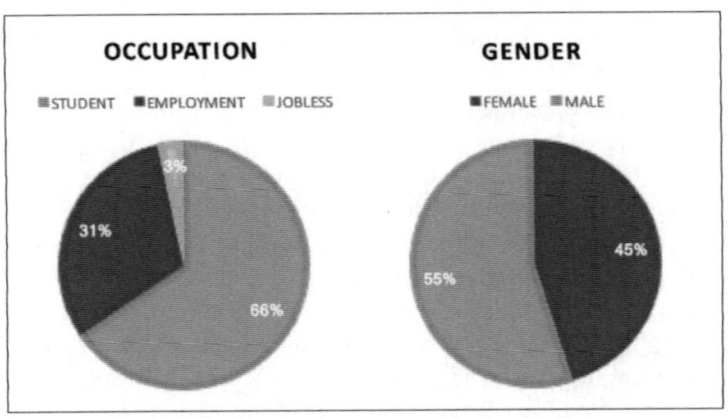

Figure 2. Occupation and Gender

A. Sleep Evaluation Questionnaire

Each participant was given a set of questionnaires to complete in their own time. Included in the set were three different questionnaires related to: sleep quality, sleep experience, and daily activities for the past month. Moreover, questions were based on different studies for diagnosing sleep disorders: Pittsburgh Quality Sleep Index (PSQI) [12], Consensus Sleep Diary (CSD) [13] and Stanford Sleeping Scale (SSS) [14]. The questions in the survey were slightly modified to be applicable to the public audience who is not familiar to mental sleep disorders [15]. In the end, the participants were asked if they would like to volunteer as subjects in the case study.

B. Case Study Procedure

Based on the results of the questionnaire study, three subjects were chosen: two males (22 and 24), and one female (26). One of the subjects (24/M), claimed to have sleep troubles consisting of: falling asleep, irregular sleeping patterns and feeling tired during the day. Au contraire, during the study, the subjects were required to keep track of: (a) food, beverages, medication, and tobacco consumed/used; (b) physical exercise (what exercise and how long); (c) activities from at least 2 hours prior going to bed; and (d) sleep evaluation after awakening. Furthermore, they did not change their sleeping schedule, consuming, and activity behavior. Before starting the experiment, the volunteers were guided on how to use the tracking tools, interviewed about their thought of the experiment, and a questionnaire related to sleep experience was given to fill in. Besides, two tracking tools were provided to register the activities: (1) a notebook, consisting of questions and feedback requests on their sleep, a blank table and some additional blank comment space; and (2) a smartphone application with questions related to their sleeping experience, tracking option for consumables and activities and a comment section (see Figure 3). Either one of the two diaries had to

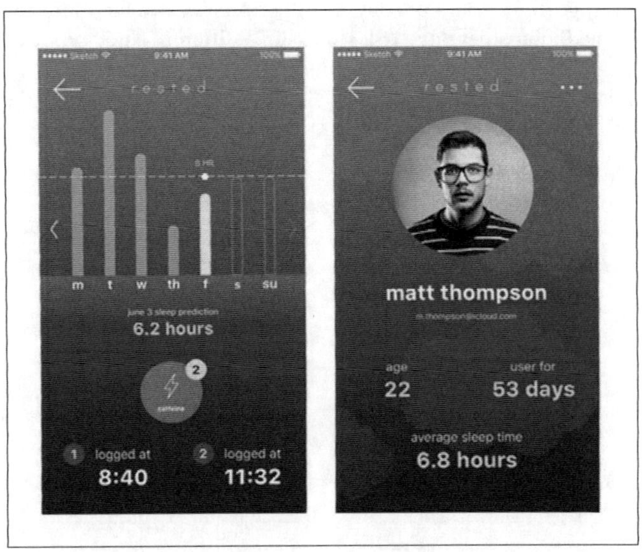

Figure 3. Tracking Application: Rested Beta 0.2

be filled. Before starting the experiment, the volunteers were guided on how to use the tracking tools, interviewed about their thought of the experiment, and a questionnaire related to sleep experience was given to fill in.

The tracking application for the smartphone was created to make tracking easier and accessible for the subjects. At the end of the experiment, the subjects were once again interviewed with the same questions as the pre-experiment interview and were given the same questionnaire about sleep experience to fill in.

The results from all data provided by the participants and subjects have been analyzed through a cause and effect study. The relationship of consuming, activity behavior, and sleeping evaluation were mapped into multiple subgroups and analyzed. The groups consisted of occupation, age, and sleep evaluation score.

III. RESULTS

The results from the sleep evaluation questionnaire are shown in Figures 4, 5, and 6. These include sleep evaluation score, bedtime, wake time, and a plot of sleep evaluation against the total amount of sleep. In this regard, sleep evaluation is based on a [0-10] point scale, where 0 is the most negative and 10 the most positive sleep experience. Participants filled the questionnaire individually, without further instructions, while questions left open were not taken into account. From the aforementioned, the participants are divided into sub-groups: (a) Good sleepers (8+), normal sleepers (5-7), bad sleepers (4-); (b) student and non-student; and (c) by age group: between 16 and 19 years old, consisting of mainly students attending high school, from 20 to 23 years old, consisting of mainly students attending university or college, and 24-27 years old, consisting of mainly working adults.

The results are compared to guidelines given on sleep for adolescents. According to previously published research, the body has optimal rest and growth when adolescents sleep on average eight hours per day. Healthy sleep is sleeping between 6 and 9 hours per day [16]. However, a different study reported that only 15% of the adolescents are sleeping eight hours [17]. Among the participants, 14,5 % (9 out of 62) did sleep the recommended 8 hours of sleep per day. In the questionnaire, certain sleep - wake behaviors were asked and are shown in Figure 5 and 6. In overall, the participants in the age group of 16-19 and 24-27 are the most satisfied with their sleep. Participants aged from 20-23 experience the worst sleep.

Group	Age	Average Sleep [hours]	S.D.	Bed time	Wake time	Rating sleep	Time Fall asleep [minutes]
All participants	22,1	6,61	0,86	00:30	07:15	6,08	22,9
Students	21,3	6,44	0,92	00:45	07:15	5,77	22,9
Non-Students	23,9	7,00	0,78	00:15	07:10	6,79	23,2
16-19	17,1	6,89	0,83	00:15	07:15	6,44	17,2
20-23	21,5	6,32	0,96	00:45	07:15	5,77	25,3
24-27	25,1	6,91	0,81	00:15	07:15	6,36	22,0
Rating <4	21,9	6,00	1,03	01:00	07:00	3,44	26,3
Rating 5-7	21,9	6,64	0,87	01:00	07:30	6,04	20,7
Rating 8>	22,6	7,11	0,78	23:45	07:00	8,50	23,6

Figure 4. Sleep Evaluation Data 1

Group	Age	% want to sleep more	% awoken by alarm	% Sleeping alone	% Screen use prior sleep
All participants	22,1	93,50%	87,10%	83,80%	96,80%
Students	21,3	97,67%	88,37%	83,72%	100,00%
Non-Students	23,9	84,21%	94,74%	84,21%	92,56%
16-19	17,1	100,00%	77,78%	88,89%	100,00%
20-23	21,5	93,55%	96,77%	90,32%	100,00%
24-27	25,1	90,91%	90,91%	72,73%	95,45%
Rating <4	21,9	100,00%	100,00%	100,00%	100,00%
Rating 5-7	21,9	89,29%	85,71%	78,57%	100,00%
Rating 8>	22,6	94,44%	88,89%	77,78%	94,44%

Figure 5. Sleep Evaluation Data 2

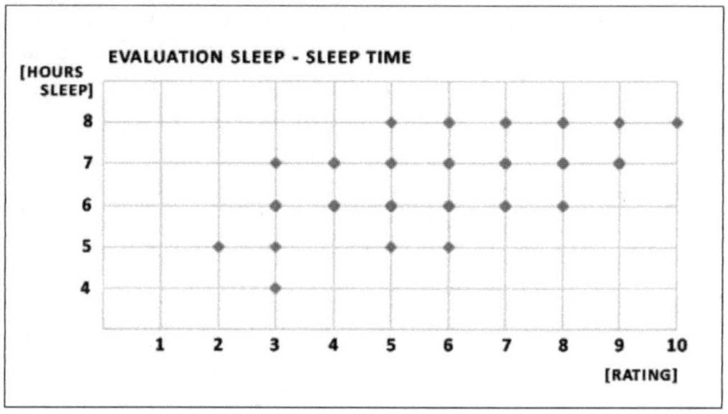

Figure 6. Sleep Time Versus Evaluation Score

The results from the case study are shown in Figure 7. A diary study gave more accurate values in sleep wake times and daily activities of the subjects. In addition, according to the data sheet, it was clear that the subjects underestimate their amount of sleep. The total amount of sleep is approximately half an hour to an hour more than the time submitted at the questionnaire. The subjects were always looking at the worst case scenario. Therefor they were underestimating their total sleep time.

Consuming and activity behavior having an impact on sleep experience does not clearly show in the results from the case study; likewise, interviews with the participants gave different answers without a clear conclusion. Notwithstanding, regular eating and drinking habits do have a small influence on one's sleep. In a different study, it was stated for optimal sleep hygiene, food, drinks, and physical exercise is not recommended prior bed time. In contradiction to the previous statement, one of the subjects (26/F) was having intense physical exercise 2-3 times per week prior going to bed. In the after-interview, she clearly stated "Physical exercise improves

	Case 1	Case 2	Case 3
Age/Gender:	22/M	24/M	26/F
Occupation:	Student	Student	Student
Overall sleep rating:	7	4	8
Average sleep:	7,5 7*	6,6 6*	7,2 7*
Weekday:	7,1	6,1	6,9
Free day:	9,75	8,1	8,5
Average bedtime:			
Weekday:	00:28	01:46	23:47
Free day:	03:13	02:32	01:20
Activities prior Bed:	Homework Watching series/movies Drinking** Music	Smartphone use Homework Drinking** Student association	Physical Exercise Showering Smartphone use
Other information:	Sleeps alone	Sleeps alone Difficulties falling asleep Very tired in morning	Sleeps with partner

*Average sleep submitted during initial questionnaire.
**Occasionally drinking alcohol [2-4x per week]

Figure 7. Case Study Data Summary

my sleep experience, it helps me to fall asleep more easily and it helps to clear my mind".

The evaluation of sleep is influenced by memories and feelings related to bad sleep. Subjects (22/M and 24/M) gave their opinion that bad sleep has more impact on their morning routine. The difference between good and bad sleep experience cannot be compared as many factors affect this judgment. The moment of evaluation is also very important. For example, evaluating your sleep directly after waking up, will result in a lower evaluation score; whereas doing it after taking a shower, or having a morning coffee, will improve the evaluation score. It has been stated by the subjects that memories of bad sleep are stronger and more memorable, while memories of good sleep are easily forgotten and replaced by new thoughts of activities and events.

On another topic, the tracking application was favored over the traditional paper notebook, during the two-week study: all three subjects preferred to use the smartphone application. One of the subjects (22/M) gave his opinion about activity tracking: "Tracking my activities gave me more insight about my daily routines and consuming behavior. However, I did not change my behavior a lot". The subjects were interested in sleep hygiene and behavior change, however, they were not willing to change their own daily routines.

IV. Discussion

The case study sampled sleep experience and activity behavior across multiple nights and outside clinical settings. The findings found in this study are based on the results provided by the participants, which are subjective statements and believes they have themselves. Moreover, the findings supported previous studies about the sleep patterns of adolescents and young adults, but a bigger sample size is required to give a final conclusion. Inevitably, the coherence between consuming and activity behavior prior going to bed is unclear and cannot be confirmed in this study. Consuming and activity behavior, a part of sleep hygiene, has on a certain degree effect on one's sleep. The positive effect has been stated in previous studies as well, but sleep hygiene is due to contradictory result material not used in this study. It can be concluded that activity and consuming behavior do have a relation to sleep experience.

When reviewing the findings and implications discussed herein, it is important to consider the limitations which may have influenced these findings. The data is in all cases subjective information provided by the participants themselves. The participants' self-evaluation is based on own knowledge and attitude towards sleep. People who are biased towards good sleeping experience, are evaluating their sleep with more care. As a result, these people were very satisfied with their own sleep.

The other group has a more diverse opinion on sleep, these individuals consisted mainly of college and university students. Hence, due to projects, exams and social relations, there is less structure in their sleep hygiene. Pressure from school, social relationships and (part-time) work is in some cases more important than sleep, which results in their sleep evaluation to be below average. Subsequently, people with a more negative view on sleep, are always looking at the worst case scenario. As a result, they were underestimating their total sleep time on a regular base. It is thought that, as a result of this study, sleep evaluation is influenced by many more factors outside of the domain of sleep. Negative emotions and memories do have a significant impact on sleep evaluation.

From the results of the sleep evaluation questionnaire, it can be argued that: (a) adolescents and young adults of this study do not reach the quota of sleep recommended for their age group, but are within margin of healthy sleep behavior; (b) subjective thoughts about sleep experiences are more negative than the actual sleep experience. 25,8 percent of the participants rate their sleep overall bad (less or equal than four on the 10-point scale), and there are only five cases where people sleep less than the norm of at least five hours; (c) most of the people are misinformed about what good quality sleep means. Their view on sleep in general very negative. This may be caused by all the information available on the Internet about various sleep disorders.

V. Conclusion and Future Works

Adolescents and young teenagers do sleep less, but within the margin for healthy sleep. However, the evaluation process of subjective sleep experience is unbalanced. Memories and emotions related to unpleasant sleeping experience are more memorable and present than pleasant sleeping experiences. Pleasant memories are more easily replaced by new thoughts and are often forgotten during the sleep evaluation process. As a result, negative emotions and memories have more influence on the results of subjective sleep evaluation.

In future studies, the domain of subjective sleep experience is not only limited to factors as overall sleep time, wake time, sleep quality, but factors as feelings, memories and emotions could be included in addition. Following the Japanese trend of design emotion, the approach of Kansei design, where research and development are focused towards the customer's psychological feelings and emotions could be used as one of the starting points.

ACKNOWLEDGMENT

The authors would like express their appreciation towards the subjects and people involved in this research, TU\E, The Netherlands for guiding and supervising the whole process, and Lee Laboratory, Tsukuba University, Japan for supervising the revision of this paper.

REFERENCES

[1] American Sleep Disorders Association, American Sleep Disorders Association. International classification of sleep disorders: Diagnostic and coding manual. Rochester, MN: American Sleep Disorders, Rochester, MN: American Sleep Disorders Association, 1991.

[2] A. L. Jr Chesson, R. A. Ferber, and J. M.Fry, "The indications for polysomnography and related procedures, " Sleep, vol. 20, no. 6, pp. 423-87, June, 1997.

[3] A. R. Wolfson and M. A. Carskadon, "Sleep Schedules and Daytime Functioning in Adolescents", Child Development, vol. 69, no. 4 pp. 875-887, August, 1998.

[4] ZubiaVeqar, M EjazHussain. "Sleep Quality Improvement and Exercise: A Review", International Journal of Scientific and Research Publications, vol. 2, no. 8, August 2012.

[5] M. A. Carskadon, et al. "Self-reports versus sleep laboratory findings in 122 drug-free subjects with complaints of chronic insomnia", Am J Psychiatry, vol 133, no 12, pp. 1382-1388, December, 1976.

[6] J. D. Edinger, A. I. Fins. "The Distribution and Clinical Significance of Sleep Time Misperceptions among Insomniacs", Sleep, vol. 18, no. 4, pp. 232-239, May, 1995.

[7] E. Libman, L. Creti, R. D. Levy, W. Brender and C. S. Fichten, "A Comparison of Reported and Recorded Sleep in Older Poor Sleepers", Journal of Clinical Geropsychology, vol. 3, no. 3, 199-211, 1997.

[8] D. Schneider-Helmert, A. Kumar, "Sleep, its subjective perception, and daytime performance in insomniacs with a pattern of alpha sleep", Biological Psychiatry. vol. 37, no 2, 859-865, January, 1995.

[9] M. H. Bonnet and D. L. Arand, "Physiological activation in patients with Sleep State Misperception", Psychosom Med. vol 5, no. 5, pp. 533-540, September-October, 1997

[10] E. Hoddes, M. Carskadon, R. Phillips, V. Zarcone, W. C. Dement , "Total sleep time in insomniacs", [Abstract] Sleep Research, vol 1, pp. 152, 1972.

[11] W. F. Seidel, et al. "Daytime alertness in relation to mood, performance, and nocturnal sleep in chronic insomniacs and noncomplaining sleepers", Sleep, vol. 7, no. 3, pp. 230-238, 1984.

[12] D. J. Buysse, C. F. Reynolds, T. H. Monk, S. R. Berman and D.J. Kupfer, "The Pittsburgh Sleep Quality Index (PSQI): A new instrument for psychiatric research and practice", Psychiatry Research, vol. 28, no. 2, pp. 193-213, May, 1989.

[13] E. Colleen, et al. "The Consensus Sleep Diary: Standardizing Prospective Sleep Self-Monitoring", Sleep, vol. 35, no. 2, pp. 287-302, February 2012.

[14] A. Shahid, K. Wilkinson, S. Marcu, C. M. Shapiro, "Stanford Sleepiness Scale (SSS)", STOP, THAT and One Hundred Other Sleep Scales, pp. 369-370, November 2011.

[15] A. J. Spielman, P. Saskin ,M. J. Thorpy, "Treatment of chronic insomniaby restriction of time in bed", Sleep, vol. 10, no. 1, pp. 45-55, February 1987.

[16] "Teens and Sleep," 2016, URL: https://sleepfoundation.org/sleep-topics/teens-and-sleep [accessed: 2017-02-27].

[17] D. F. Kripke, et al. "Mortality Associated With Sleep Duration and Insomnia", Arch Gen Psychiatry, vol. 59, no. 2, pp. 131-136, February 2002.

Study on the Effectiveness of Coloring on Physical and Mental Health of the Elderly in Adult Daycare Centers

Chuan Feng*, Ya Wen Deng†, Naomi Okubo‡, Tetsuya Sanpei§ and SeungHee Lee¶
*Kansei, Behavioral and Brain Sciences – Graduate School of Comprehensive Human Sciences
University of Tsukuba, Ibaraki, Japan
Email: feng111@outlook.com
†National Tsing Hua University – Institute of Service Science, Hsinchu City, Taiwan
Email: dengyawenjp@gmail.com
‡coco de class Corp., Ibaraki, Japan
Email: n.okubo@cocodeclass-afs.com
§Japan Prime Computing Corp., Ibaraki, Japan
Email: sanpei.tetsuya@j-pc.jp
*Faculty of Art and Design – Graduate School of Comprehensive Human Sciences – University of Tsukuba, Ibaraki, Japan
Email: seungheekansei@gmail.com

***Abstract*—It has been reported that coloring provides a wide-range of health benefits for all ages, especially the elderly. Many nursing homes started to apply coloring as a routine activity for their residents in Japan, and its usefulness is expected. As the rapidly declining birthrate and aging of the population continue, it is necessary to develop new activities to improve the lifestyle of senior adults. The purpose of this study is to assess the effectiveness of coloring therapy for the elderly on their mood states, and to investigate whether there is a correlation between color and the mood stage while coloring. A daily experimental scheme is proposed for the elderly in a daycare center by providing purposeful line drawings with concrete elements.**

Keywords–coloring; art therapy; elderly; daycare center.

I. Introduction

As Japan enters the super-aging stage, the number of people certified for long-term care/support has been increasing. Citizens over the age of 65 account for 26.7% of the country's population, and only around 1.7 million care workers are available for almost 6.2 million individuals requiring special attention [1]. This increase in elderly per caregiver ratio results in a heavy burden for the worker and a reduction in quality care. In response to this, many nursing homes arrange activities beyond daily care to lessen stress on caregivers and provide a better living environment.

Recently, a number of nursing centers started applying 'coloring' as a routine activity, since it has been reported that coloring books have a wide-range of therapeutic and health benefits [2]. Furthermore, related research proved that the use of coloring in non-pharmacological treatment made a positive impact on patients with various diseases of aging, like cognitive disorders and disabilities. Hattori et al. [3] found significant improvements in the vitality and quality of life on patients with mild Alzheimers disease. Moreover, it is suggested to be an effective stress reducer for persons with intellectual disability [4], while cardiac rehabilitation patients also experienced a greater relaxation and positive mood response after the coloring therapy [5]. However, although the use of coloring books is encouraged as a pleasurable activity, it is not considered as professional art therapy [6].

In this regard, the purpose of this study is to determine the effectiveness of coloring on physical and mental health of adult daycare center users, and provide evidence on whether or not this activity affects the elderly. This paper is structured as follows. In Section I, the current state of Japanese elderly care and coloring as a therapy is introduced. In Section II, general information of the cooperative facility is given, and in Section III and IV, the methodology and preliminary study are described, respectively. Finally, a brief discussion and conclusion for the current study progress are given in Sections V and VI.

II. Daycare Center

For this research, a daycare center in Ibaraki, Japan, was visited. This facility nurses 32 senior adults in need of Long-Term Care(LTC) / Support, 11 males and 21 females, ranging between the age of 60 and 105, with a mean age of 87. According to the Japanese Ministry of Health, Labor and Welfare, the level of care is classified from Support Level 1(Needs social support and partial assistance with ADLs but may maintain or improve functions), to Care Level 5 (maximum demand for full service) [7]. Following this classification, the users of the

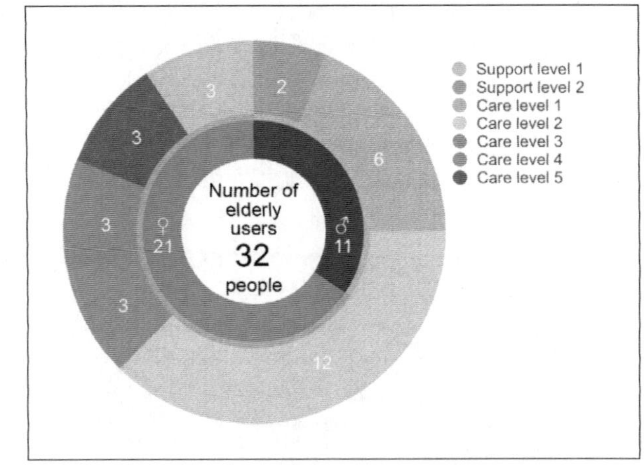

Figure 1. Number of Elderly Users and Care Levels

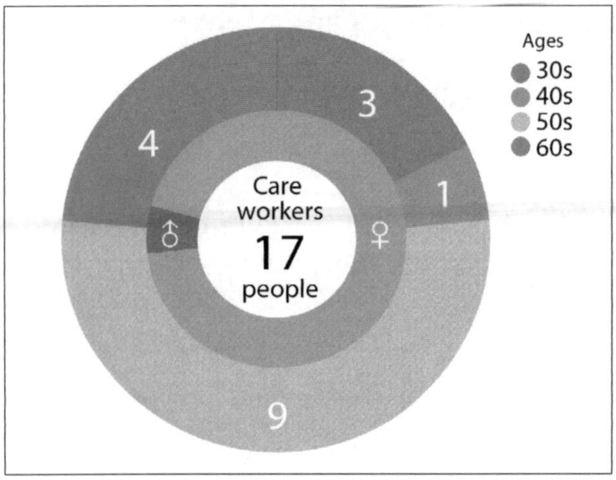

Figure 2. Number, Ages and Gender of Care Workers

care center are distributed as presented in Figure 1, while 81% of the elderly able to do coloring are below Care Level 3.

On the other hand, Figure 2 shows that from the 17 care workers at the daycare center, only one is male. What is more, the majority of carers are concentrated on the 50-60 age range group, while no one is in their twenties.

III. Methodology

This study included a total of N=30 elderly. Even though most of the participants are under Care Level 3, they are more or less suffering some common geriatric disease, such as: hypertension, heart diseases, bone and joint pain or diabetes. In addition, some of them are observed to have different degrees of progressive memory loss and cognitive impairment, namely suspected of having Alzheimer's disease (AD). For this reason, in order to avoid causing discomfort, the length of each coloring time and the complexity of the drawings were designed as stated below.

A. *Drawing Design*

Considering cultural influences, Japanese elderly prefer contents which are familiar and related to their past. To elicit emotion and empathy, three themes were chosen: 'Flowers' and 'Traditional Objects' (Figure 3). The contents of these line drawings are colorless and were previously tested with the elderly to not be difficult for the subjects.

B. *Procedure*

Firstly, the participants, after signing an informed consent, are presented with a set of 12 pre-determined colored pencils and line drawings. Coloring time is divided into two sessions of 30 minutes each. On the first session, the subjects will color an specific drawing, whereas, on the second session, the elderly are free to choose which drawing to color.

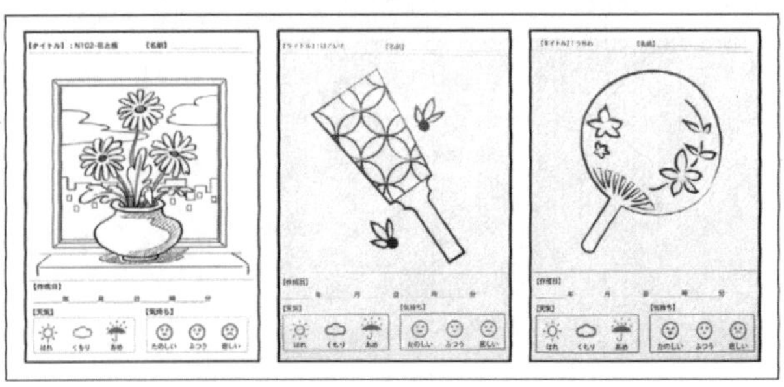

Figure 3. A Sample Drawing From Each Theme

Figure 4. Answer Icons

Secondly, before starting coloring, the subjects are asked to record the 'date', 'weather', and 'mood' at the time of the activity. To account for participants with mild cognitive impairment, each answer uses easy to understand icons as shown in Figure 4.

Finally, the finished coloring works are collected and scanned to build a digital database. Then, the 'color', 'drawing', 'blood pressure', 'hearbeat rate', 'body temperature', 'date', 'weather', and 'mood' elements are analyzed to find relationships between the activity and change on the elderly. Furthermore, a color analysis program will assess color usage, line pressure, and line consistency of each work.

IV. Preliminary Study

For the purpose of collecting design data, a preliminary study was conducted at the daycare center. A pilot study with prototype drawings (Figure 5) showed that most of the elderly were willing to try coloring. During the activity time, the atmosphere became lively and relaxed. In addition, two opposite phenomena were perceived: some of the subjects were extremely devoted and refused to communicate with the outside world; while some actively communicated with people, showing a significant social desire.

Additionally, Figure 6 presents the coloring works of one subject (age 92). This female participant is a farmer living with her son and daughter in law, husband deceased. According to the information provided by the daycare center, she has been attending this institution for 3 years, 6 times per week, and suffers from astigmatism and arrhythmia. On the other hand, during the activity, she showed careful thinking while filling the line drawings. Moreover, some of her personality traits made detectable during the experiment were: will not say anything when depressed, murmurs while drawing if feeling angry, gets easily angry when remembering bad experiences, and goes along with other people.

Figure 5. Pilot Study on Coloring

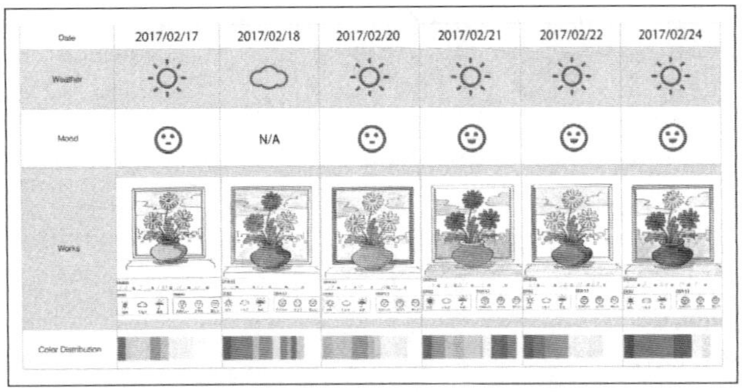

Figure 6. Collected Data from a Subject

V. DISCUSSION

This study found that the following points need more strictly control for the upcoming experiment. On one side, there are many factors of external interferences on the processes of coloring and not all of the elderly can finish their works everytime. On the other side, the pilot study did not found any significant relationship between the emotion of subject and the color they have used. Therefore, in the upcoming experiment, the study will introduce the experiment more specifically to the elderly, and briefly interview them about their feeling while coloring to have more information to analyze.

VI. CONCLUSION

The purpose of this study was to determine the effectiveness of coloring as a therapy for elderly nursing home users, and provide evidence supporting the positive effects that coloring can make on mental and physical health. A daily experimental scheme was proposed by providing meaningful line drawings with figurative elements to the elderly, to collect, analyze and track their daily mood.

From the guidelines presented in this paper, a three months research will be conducted at the daycare center to gather and process the data.

ACKNOWLEDGMENT

The authors would like to thank all colleagues and students who contributed to this study, and are grateful to the National Federation of Agricultural Cooperative Associations, the daycare center of JA-Hitachi Fureai, Tokai, and the daycare center of JA-Hitachi Daisuki, Takahagi, for their collaboration during preliminary investigations.

REFERENCES

[1] "Considering a Social Model to Overcome Population Aging," Japan Ministry of Health, Labor and Welfare, Tech. Rep., 2016.

[2] "Why grown-ups love coloring books too," 2016, URL: http://ideas.ted.com/why-grown-ups-love-coloring-books-too/ [accessed: 2016-06-26].

[3] H. Hattori, C. Hattori, C. Hokao, K. Mizushima, and T. Mase, "Controlled study on the cognitive and psychological effect of coloring and drawing in mild Alzheimer's disease patients," Geriatrics Gerontology International, vol. 11, no. 4, 2011, pp. 431–437.

[4] C. Schrade, L. Tronsky, and H. Kaiser, "Physiological effects of mandala making in adults with intellectual disability," The Arts in Psychotherapy, vol. 38, no. 2, 2011, pp. 109–113.

[5] D. Elkis-Abuhoff, M. Gaydos, R. Goldblatt, M. Chen, and S. Rose, "Mandala drawings as an assessment tool for women with breast cancer," The Arts in Psychotherapy, vol. 36, no. 4, 2009, pp. 231–238.

[6] R. Carolan and D. Betts, "The adult coloring book phenomenon," 2015, URL: http://3blmedia.com/News/Adult-Coloring-Book-Phenomenon/ [accessed: 2016-06-26].

[7] "About the Certification of Needed Long-Term Care," 2015, URL: http://www.mhlw.go.jp/topics/kaigo/kentou/15kourei/sankou3.html [accessed: 2016-09-05].

A Diary Study of Smart Phone Notification System For College Students

Ji Wang*, Qingyao Zhang, Qinqing Gong†, SeungHee Lee*, §
*Kansei,Behavioral and Brain Sciences
Graduate School of Comprehensive Human Sciences
University of Tsukuba, Ibaraki, Japan, 305 0005
Email: wangji@kansei.tsukuba.ac. seungheekansei@gmail.com
†School of Psychology
Shaanxi Normal University, Xi'an, China, 710062
Email: qingyaozhang@outlook.com purivy0430@sina.com

***Abstract*—Smart phone can support a variety of functions in people's daily life. Notification is one of the basic core function of smart phone user experience. This research explored the college students ' behaviors when they deal with the notification system of smart phone. In order to know the behaviors of user, this study conducted a diary study method to record the information about notification users received from the smart phone. Before the one week diary study, a pre-experiment interview was conducted, the results of interview shown that they have a negative evaluation to receive notifications (disruptive). Hierarchical linear model analyses on the diary contents suggested that the current task types had significant influence on the college students' evaluation on notification disruptive. The evaluation on the importance of notification content can predict user's evaluate on the disruptive of notification content.**

Keywords*—*notification system; smart phone; dairy study; user experience.

I. Introduction

Smart phones have become an integral part of college life, current smart phone can support a variety of functions. Notification is one of the basic core functions of smart phone. Notification is used to drive the users attention to the content of a notification message and remind user about something. As the media for user to communicate with the outside world, it is difficult to measure the importance of notification. However, many users feel pressure for the notification system. Previous studies have shown a core problem with notifications was that users tend to drop their current task to check the notification instead [1]. In the work context, this can be explained by the fact that notifications interrupt workers. Information workers find it difficult to return to a previous task after having been interrupted by a notification. They report higher subjective workloads when receiving notifications during phases of focused work. However, previous studies have also shown that people value notifications. Those studies also suggested that notifications vary in different levels of importance [2]. A diary study was conducted to explore the extent to what kinds of notifications are needed by college students. Dairy method have high ecological value as they carried out in the users real environments [3]. It helps the researchers to get in-depth understanding of users day to day jobs and environment diary study can be used which involves participants reporting their activities over a specific period of time in their normal daily lives addressing the when, the what and the how. This study will use diary method, which will record daily events of the user deal with the notifications accurately. It helps to understand how the notification usage changes over the time (Compared to a traditional in lab usability study where this study typically gather information on firstly time usage of a system or product) and the kinds, frequency of notification and so on.

The structure of the paper is as follows: Section I introduces the essence of the smart phone and notification system. In section II the method of the experiment is introduced. Section III reports the result of the diary study on the smart phone notification system. In Section IV, the analysis and discussion of the diary study results are presented. Finally, in Section V, the conclusions of this research are laid out.

II. Method

A. The Participants

Fourteen participants from Shannxi Normal University took part in this experiment , 8 female and 6 male, with the age distribution 18-24 years old (M = 20, SD = 1.96). 2 out of 14 participants (one female and one male) interrupted the diary collected during the study,so this study totally collected twelve participants' data.

B. Device

All the mobile phone (provided by participants) used in this experiment with the version above Android OS 4.3 or iOS 8.1). Participants were required to install Smartisan Notes (A notetaking app.User can add picture in the notes and download this APP in Apple Store or Google Play for free). Participants should set on their phone notifications system in the mode that the notifications come and it can show on the lock screen.

C. Procedure

Pre-experiment interview was conducted to know the background information about participants smart phone notification interview questions included :

Q1: How many app installed on your phone?

Q2: These types of app (social network/ video and image/education/payment/online shopping service), how many of each kind app and how you set the notification systems of these app ?

Q3: Share a recent experience about the Notification affect (disturb) your study or work ?

Dairy study:

Contents of dairy collection as shown in Figure 1

Date	Time	Current task	Location	Source of information	Notification context	Grading importance 1 2 3 4 5	Grading disruptive 1 2 3 4 5
2016.9.7	08:13	Reading book	Library	WeChat	News feeds		
	09:25	Running	Playground	QQ	Group chatting		
	13:16	Sleeping	Room	Phone	Package		

Figure 1. The Example of Dairy Content

Each participants took the records of dairy content for one week [2].

III. RESULT

In this experiment, pre-experiment interview were conducted for 14 college students, they installed 11 to 42 applications (M = 23.64 , SD = 9.39) in their smart phone. These applications can receive notifications when connected network. Based on the previous survey of applications types, we classify the applications into: social networks, video and images education, entertainment, payment and life services.

We collected a total 739 notification messages information from 12 participants in one week. Coding scheme: gender codes are: 0 = female, 1 = male; grade codes are: 1 = freshman, 2 = sophomore, 3 = junior students, 4 = senior students.

There are 21 categories in current task, 9 categories in location and 21 categories in notification sources. Based on the real life situation (classification of apple store applications), we coded the current task again. To facilitate the analysis of the hierarchical linear model, we did the dummy coding for current task, location, notification sources. Current task was coded as 4 dummy variables location was coded as 3 dummy variables, the notification sources was coded as 4 dummy variables.

To test the the relationship between notification sources and the importance of the notification (short for importance) , we built the multilevel random slopes model in this study. Results showed that the mean value at level 2 of dummy variables entertainment, life service, and education achieved significant difference ($P < 0.05$) , indicated that the source of the notification message affects participants evaluation about the importance and urgency of a specific message. Specifically, compared with the notification from the social networks, the notification from entertainment and eduction applications were considered less important, life services application notifications were more important. Additionally, there was no significant difference between the importance of notification from social networks and life services applications. The result of primary data was shown in Figure 2. It shows the average rating of disruptive at different times of the whole day. Morning and afternoon periods were more disruptive than other periods, because it was the study time for college students.

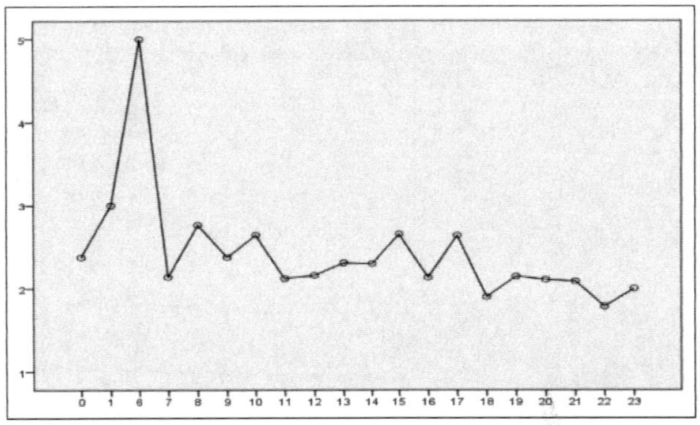

Figure 2. The Average Rating of Disruptive at Different Times

IV. DISCUSSION

From the dairy content, we found that college students interact with their smart phone all the time in their daily life. They basically put attention to the notifications when they came. This finding is consistent with previous study[1], even though the importance of those notification content were not high. Hierarchical linear model analyses was conducted on the diary contents, the results shown the current task types had significant influence on the college students evaluation on notification disruptive. When college students focused on current tasks, especially those tasks correlated with study, the interrupt caused by dealing with notification make college students rated the disruptive of notification in a high score. This study also found he change in on the importance of notification content can predict users evaluate on the disruptive of notification content. The results of this study can give suggestion to the college students to deal with the notification system. For instance, when students are carrying out study or work task they can choose to set the notifications system to Do not disturb or Mute mode, it will reduce the times of notification interrupt the current task.

V. CONCLUSION

This research carried out a dairy study to explore college students evaluations about the notifications of their smart phones. It suggests a new usage model for smart phone manufacturers to develop in their devices for college students. The notification sources were categorized into four categories, they were, listed in descending order of importance, life services, social networks, entertainment and education. Additionally, the current activity affects participants' evaluation about the interference of the notifications. Specifically, when they were studying or working, they considered the notifications are more noisy. Due to the limitation of sample size, a wider range of participants–including office staff, programmers and other brain-workers, could yield an opportunity to more broadly generalize the results.

REFERENCE

[1] M. Czerwinski, E. Horvitz, and S. Wilhite , A diary study of task switching and interruptions, in Proceedings of the SIGCHI conference on Human factors in computing systems. ACM, 2004, pp. 175-182, 2004.

[2] A. Sgirazi et al. , Large-Scale Assessment of Mobile Notification, in proceedings of the SIGCHI conference on Human factors in computing systems. ACM, 2014, pp. 3055-3064, 2014.

[3] A. Sigh and S. Makhorta, A researchers guide to running diary studies, in proceedings of APCHI 2013, pp. 296-300, 2013.

Kid's Friendly Wearable Device for Children's Daily Safety

SeungHee Lee
Faculty of Art and Design
Graduate School of Comprehensive Human Sciences
University of Tsukuba, Tsukuba, Japan
e-mail: seungheekansei@gmail.com

Abstract— **This study introduces the development of a practical product based on Kansei simulation of children's behavior and physiological reaction against anxiety. It started in 2006 with children under 6-years-old who had few experiences of accidents or events, and less linguistic communication skills with their parents. The reliability of heart rates to detect fear and anxiety was tested to know if it could be one of the functions on this device. The heart rates and behaviors of children while watching simulated movies showing dangerous situations was analyzed. With the support of the Ministry of Internal Affairs and Communications, a device named "Onigiri Machine" was developed to keep children safe while their parents work during daytime. It was highly evaluated by scientists worldwide through conferences papers, newspapers, and magazines. Now, keeping track of children and elderly in shopping malls or in hospitals became an issue in Tsukuba, Japan. For this reason, the device was refined to update the functions and qualities for practical uses in public services. Finally, this study shows how the device can be used to keep children and elders safe in public spaces, and how it delivers personal based data in an easy interface for their companions.**

Keywords- wearable device: kid's friendly design; information sharing; children's safety.

I. INTRODUCTION

The idea of this study was based on personal interests in security for children between 0 to 6 years old in nurseries or kindergartens while parents work during daytime. For the development of the device, a basic research on physiological data from children between 1 to 6 years was needed to detect unconscious states in children through their heart rate.

Generally, regular working mothers should stay with their children at least for 8 hours a day, and should communicate with them, face to face, for 4 - 5 hours a day in average. It is hard to say that this is enough to know what is happening to their children's growth, both physically and mentally. Furthermore, the linguistic communication ability of 0 to 6 years old children, before elementary school, is not enough to be correctly understood by adults. In this regard, the author had some frustration to communicate and get information on her child, meaning that the only way to understand was to keep watching her behavior and reactions during limited moments. On the other hand, children under 6 years old get most of their influences from people around them while imitating their behaviors naturally. As such, there must be a clever way for working parents to know and feel their children.

This study introduces the development process from researching on children's behavior and emotional reactions to the final stage of making a practical system for real users in public places. In order to develop a device that can detect children's conditions in anxious or emergency situations, the following components were selected: heart rate measurement device, GPS, camera, 3D accelerometer, XBee network, and a microcontroller for connecting and controlling the sensors. The most outstanding progress of this study was done between 2009 and 2011, with the aid of the Strategic Information and Communications R&D Promotion Program (SCOPE) funded by the Ministry of Internal Affairs and Communications. Based on the children's heart rates experiments started in 2006, a device named "Onigiri Machine" was developed and updated to support information sharing of children and elderly who cannot communicate satisfactorily with people.

The structure of the paper is as follows. In Section II, the basic research used as a base for this study is introduced. In Section III, a design approach based on behavioral aspects of children concerning safe construction of form, namely "Kid's Friendly Design", is presented alongside advanced technologies to detect a variety of situations. Section IV details the experiment conducted for this study. Finally, Section V concludes this research.

II. BASIC RESEARCH

This study uses as base a foundation research done in 2006 concerning heart rate measurements on children during stressful situations combined with the Kansei approach of design.

A. Kansei

Kansei is a Japanese term that implies subjective perception and reaction based on experiences. It involves emotion, preferences, and learning of daily habits. People react to an event depending on their personal experience, including knowledge and intuitive response from image matching of the past. Kansei reaction is more important for surveying behavioral habits, in addition to being more creative in finding new solutions for design development.

B. Heart rate Experiments

Heart rate is one of the elements in measuring unconscious feelings of anxiety while facing new

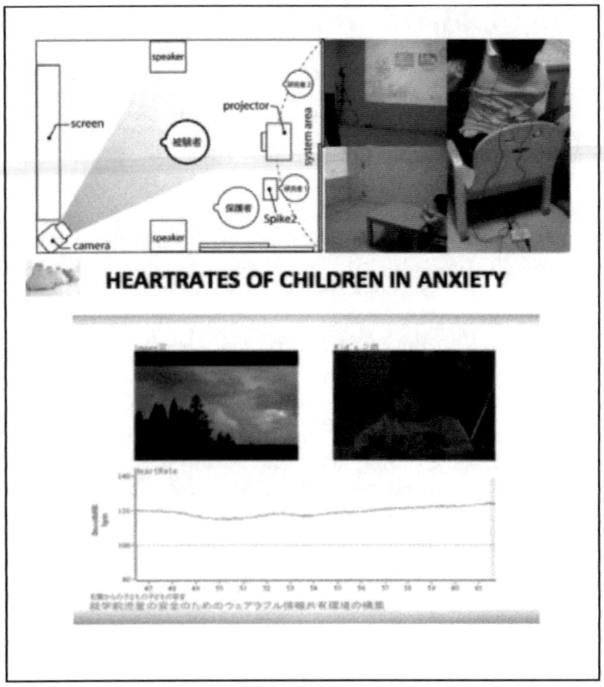

Figure 1. Physiological test of anxiety for Children under 6 years old

experiences or any uncomfortable situation. The purpose of the experiment was to know how heart rate relates to emotional prediction in children while the device detects their conditions during any situation. Figure 1 shows the setup for the experiment. Children in a nursery from 1 to 6 years old participated. Simulation films which entail dangerous and anxious situations from an adult standpoint were prepared. These movies were originally edited with royalty free copies. All the participants' parents agreed with the Institutional Review Board (IRB) of University of Tsukuba while the safety of the experiments for testing the heart rates was thoroughly checked. The result of the experiment was that children's Kansei is generally expressed around 2 - 3 years old with practical experiences and learning from their parents or people around. Therefore, their emotional reaction gets closer to an adult's.

III. Design Development

For the development of the "Onigiri Machine", a unique approach was proposed. To make children attracted to, and willing to, wear the device, a fascinating shape was designed before deciding on the technological components. As shown on Figure 2, this approach was denominated "Kid's Friendly Design".

A. Maintaining the Integrity of the Specifications

For "Kid's Friendly Design", the first step was to decide on the weight and size of the device, followed by how and where to wear on a child's body. Graphical design and decoration was the last step. The main goal is to create a fascinating appearance for kids to keep wearing it while the necessary information is read from their chest. Emotional and cognitive behavior could be detected by 3D acceleration and heart rate measurement, in order to share their status with their parents online, complemented with an image from the child's point of view.

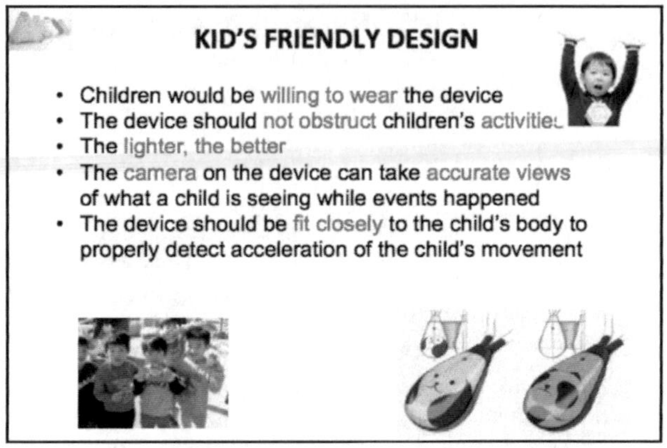

Figure 2. Kid's Friendly Design Approach

B. Weight and Size of the Device

In accordance to the Research Institute of Human Engineering for Quality Life of Japan, which has a database of the average body size for Japanese children between 0 and 12 years old since 2005, the size of the device was decided. The width of a 3-year-old child's chest is 164mm and for a 5-year-old is 176mm. Based on this information, the device should not exceeded 100mm in diameter. Moreover, most of the objects that can be attached on child's neck weight between 30 to 50 grams, while on their waist are less than 100 grams. Additionally, a survey among mothers in the kindergarten reveled that 75 percent of them would allow devices under 100 grams to be attached to their children. For reference, an alarm buzzer weights between 45 to 60 grams, likewise mobile phones for children range between 130 and 150 grams.

C. How and Where to Wear the Device

To attach the device on an appropriate position on

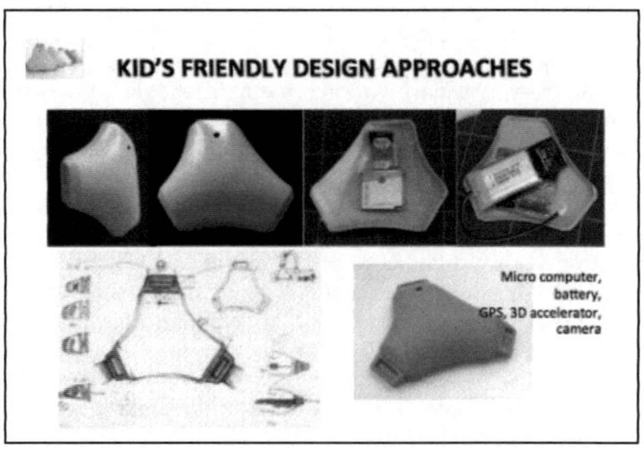

Figure 3. Kid's Friendly Design Approach

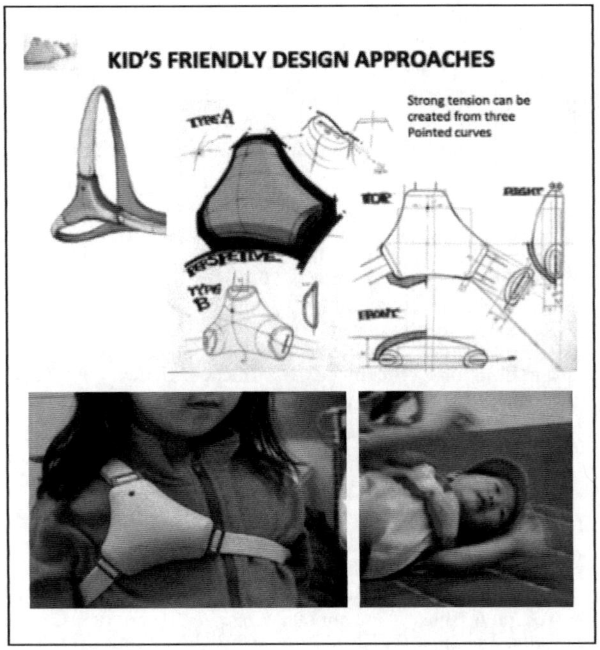

Figure 4. Implementation of Onigiri Machine

children's body, a survey of the sizes of desks and tables in the kindergarten was conducted. The machine should not disturb the activities of the kids in the classroom or in the playground. At the same time, the camera on the device should be set at a stable position on the chest, so as to allow them to actively run around.

D. Building the Prototype

Based on shoulder sac types of design, the built of the sensors followed the concept "B-3", presented in Figure 3. The total weight of the built-in sensors had to be at most 90 grams. Notwithstanding, a necklace type of design was rejected from the candidates, due to the burden generated on the children's neck. From a pool of four ideas for shoulder types, "B-3" offers a relatively high position for the camera, while being easy to wear. An updated sketch of "B-3" was prototyped using a 3D printer to verify the size of the device. Furthermore, the layout of the sensors was tested. What is more, small connection parts were created on the corners for the straps attached to the body. Finally, an on-off switch was placed on the rear side in a tiny hole, as to be securely turned with an originally designed stick, thus preventing unwanted manipulation.

Figure 4 shows the final device 'Omusubi' (=Onigiri), which means 'rice ball' or 'connection' in Japanese. A 'rice ball' is a traditional snack between lunch and dinner popular among all ages in Japan. 'Connection' is also a meaningful concept within this study, as information is not only shared between children and parents, but also with local caretakers.

IV. Experiment

For the experiment, 12 hand-made devices were built with a setup of over 30 relay sensors to monitor the position of the devices every 10 meters inside the kindergarten. For the final user test, 50 children were selected through an ethical research pre-survey with their parents. The subjects were in average ten children per age. To observe the behavior, and check both physical movements and digital data of the children with the devices, 5 selected personnel attended the test.

Complementary to the device, a website for parents, shown in Figure 5, was made available at the beginning of the experiment. With this tool, they could monitor the movement of their children inside the kindergarten, besides watching their point of view constantly. Nevertheless, for everyday use, these shots are meant to be taken when a child registers a sudden change in heart rate or shifts his/her position up and down rapidly.

A. Extension of the Experiment

One of the biggest shopping malls in southern east Japan adapted this system to prevent children getting lost while shopping in the facility. In addition, with the function of measuring heart rates, shops can get information on when the children got excited in relation to certain products, making this feedback useful for the store to promote their products or events practically. On the other side, for detecting the position of lost children, a GPS was implemented in the device. However, as indoor measurements were not accurate, and XBee mobile network was deployed at the shopping mall for every 30 meters for monitoring and controlling the devices. Figure 6 shows the setup and positioning of the XBee mobile network inside the mall.

V. Conclusion

To develop a new device concerning children's safety, emotional and physical growth with new experiences are the most important points. Those aspects were treated using an approach with Kansei, physiological data, and behavioral movements sensed by highly technical solutions. For practical uses of the device, fascinating and reasonable design, fitting children's preferences, was highly evaluated. Currently, real production of the "Onigiri Machine" is being prepared based on the behaviors and preferences of children and elderly. In this regard, it will increase its potential: not only as a tool to prevent lost people, but also as a mean to

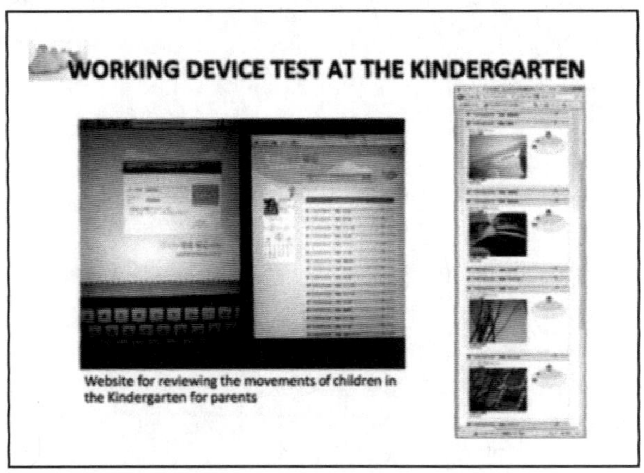

Figure 5. Online Website Interface for Onigiri Machine

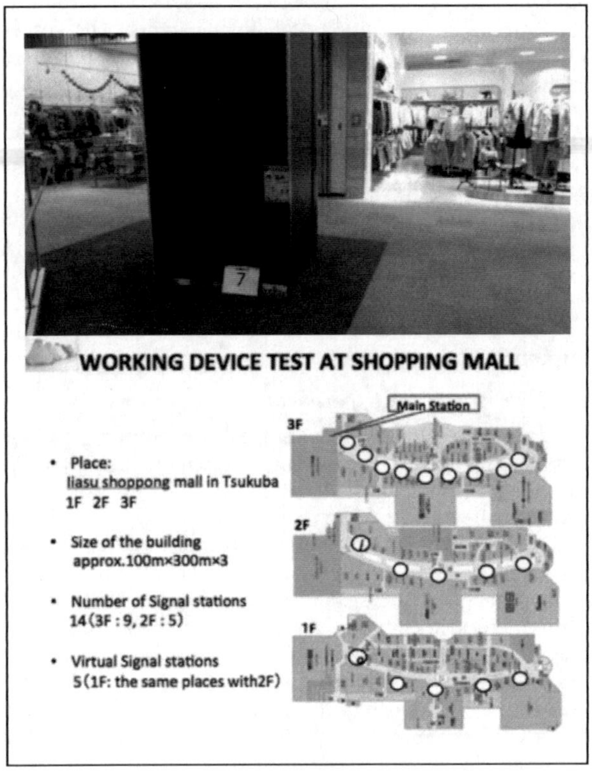

Figure 6. Signal receptors position in the mall

know the user's Kansei in various situations, regardless of age. With the mechanical engineering and information technology developed, this concept, that started from a personal need to prevent loss or kidnaping of children, brought more possibilities for the improvement of Information and Communication Technology for the future.

REFERENCES

[1] S.H.Lee, J.H.Sohn, A.Usami, M.Hamanaka, Development of Wearable Device by Kid's friendly Design for Kid's Safety, Human Computer Interaction 2010, Brisbane, Australia, 2010.

[2] S.H.Lee, M. Hamanaka, Y.Iwamoto, D. Kim, Heart-rate Response to Simulated Anxious Events- Development of Kid's Friendly wearable device for children's safety, Design and Emotion Conference 2010, Chicago, USA.

[3] M. Hamanaka, Y. Murakami, A. Usami, Y. Miura, S.H.Lee, System for Detecting Kindergartners'Potential Emergency Situations, World Multi- Conference on Systemics, Cybernetics and Informatics 2010(WMSCI2010), July Florida, USA, Vol. I, pp. 296-301.

[4] J. Farringdon, A.J.Moore, N.Tilbury, J.Church, P.D. Biemond, ,Wearable Sensor Badge and Sensor Jacket for Context Awareness. ISWC, p.107. 2009.

[5] N. Kern, B.Shiele, H. Junker, P. Lukowicz, G.Troster, Wearable sensing to annotate meeting recordings Euro-Par 2006. LNCS, vol.7, No.5, pp. 263-274. Springer, Heidelberg London. 2003.

[6] M. Wiklund, Usability in practice: how companies develop user-friendly products. Academic Press Professional, Inc., 1994.

[7] "Japan's already-safe streets could become even safer for children", 2010, URL: http://www.csmonitor.com/World /Global-News/2010/1201/Japan-s-already-safe-streets-could-become-even-safer-for-children/ [accessed: 2017-02-26].

Improving Diagnosis-related Groups with a Computerized Clinical Coding Tool for ICD-9-CM Codification

Cecília Coimbra, Marisa Esteves, Filipe Miranda, Filipe Portela, Manuel Filipe Santos, José Machado * and António Abelha
Algoritmi Research Centre,
University of Minho
Braga, Portugal
e-mail: {id6881, id6884, id6883}@alunos.uminho.pt, {cfp,mfs}@dsi.uminho.pt {jmac, abelha}@di.uminho.pt
* Corresponding author

***Abstract*—In recent years, in *Centro Hospitalar do Porto* (CHP), a major Portuguese hospital at the North of the country, there has been felt an increasing need for a computerized clinical coding tool to aid in the codification of the episodes of hospital discharges from patients admitted to its healthcare units. The process was slow and performed manually by the coding professionals, not having a centralization and unification of the information and processes associated with the clinical coding of a hospital discharge. Hereupon, in the context of this study, the aim of the present work was to design and develop a clinical coding tool for International Classification of Diseases, Ninth Revision, Clinical Modification (ICD-9-CM) to support the clinical practice in healthcare units. The coding of hospital discharge processes enables the grouping of episodes into diagnosis-related groups (DRGs) that correspond to clinically coherent and similar groups that are expected to use the same level of hospital resources. The main motivation for the implementation of this classification system is that it provides a financial and patient classification system, trying to contain the costs and waste associated with healthcare services. Thereby, a single-page application (SPA) for ICD-9-CM was designed in order to help health professionals of CHP in their daily work, namely the clinical coding of the episodes of hospital discharges. The main advantages and contributions of the development and use of this Web application are the centralization of information and tasks associated with the coding of hospital discharges, the increase of productivity and the reduction of wastes of time. Consequently, the ambition is sought to mainly improve the quantity and the quality of work performed by coding professionals.**

Keywords—ICD-9-CM; Diagnosis-related Groups; Hospital Discharges; Single-page Application; Proof of Concept.

I. INTRODUCTION

The health sector represents a tricky situation for the people and systems responsible for the storage and processing of that kind of information. The problem is not in the lack of data but in the diversity and complexity of the health field. An hospital offers a wide range of services for each patient per clinical condition leading to hundreds or even thousands of specific and unique situations. It is then of extreme importance to find a way to measure the hospital productivity and aggregate the multiple activities performed at a healthcare unit. This led to the adaptation of the DRGs to this scenario. These kind of classification systems rely on a prior coding system that translates all the diagnostics, procedures, external causes and morphologies into universal codes like ICD-9-CM, increasing the semantic interoperability and reducing the ambiguity of a discharge report. The introduction of new Web frameworks and solutions for Web development resulted in a new wave of codification platforms for this kind of code systems. With modern layout and intelligent helping tools, it is possible to reduce the codification errors dramatically and increasing efficiency.

This paper presents an insight into the development, implementation and impact analysis of a Web application directed to the ICD-9-CM codification in a major Portuguese hospital in the North of the country – CHP. The focus of this paper is mainly to highlight the differences between the prior method (by hand) and the new one.

In Section II, the state of the art and similar works are briefly described. Thereafter, in Section III, the research methodologies adopted are presented. Section IV – "Single-page Application for ICD-9-CM Codification" – presents the work developed and its main results, followed by a brief Strengths Weaknesses Opportunities and Threats (SWOT) analysis in Section V. In Section VI, the conclusion and future work concluded this paper.

II. STATE OF THE ART

The present section intends to highlight the main topics addressed throughout the work as well as main studies from the scientific community regarding medical codification.

A. *Diagnosis-related Groups*

DRGs are a classification system for inpatients at acute healthcare units aggregating under the same code cases with similar resource consumption and clinical similarities [1][2]. Developed at the Yale University (U.S.A) in the 60's, it is used since 1983 by Medicare to calculate the compensation in cases of hospitalization [3]. Due to the immense success on that field, the system was adopted throughout the world in healthcare systems to improve the planning and allocation of funds reducing the disparities and the errors.

The DRG requires a minimal dataset (MDS) in order to attribute one of the 25 main diagnosis categories to the discharge report [1]. So, the MDS includes, as follows [4]:

- The main diagnosis responsible for the patient admission;
- Other diagnosis;

- Procedures performed on the patient during the internment;
- Gender, age and height;
- Destination after discharge (transferred, death or discharged against medical order).

Each DRG group has an associated relative weight and weighting coefficient, as well as an exception threshold for the number of hospitalization days that helps convert each case into equivalent patients [4].

Wilm Quentin and colleagues in "Hospital Payment Based on Diagnosis-related Groups Differs in Europe and Holds Lessons for the United States" highlights the differences between the original DRG and the one that countries like France, England or even Portugal implement. This adaption is the basis of most European countries method to finance hospitals proving to be less cost worthy with high quality of services [2].

Carina Fourie et. all present in "Systematically Evaluating the Impact of Diagnosis-related Groups on Healthcare Delivery: A Matrix of Ethical Implications" a study of ethical implications and importance of the DRGs in diverse Swiss hospitals [3]. On other hand, in order to highlight the diversity of the subject, Yantao Xin presented a comparison of the amount of medical waste generated in major healthcare units using as basis the DRGs [5].

B. ICD-9-CM

Every time a patient is discharged from a healthcare unit a discharge report is issued with daily logs from the physician in charge of that patient. The ICD-9-CM is a perfect fit to encode diagnostics, medical procedures, external causes and morphologies, consisting in a list of codes recognized anywhere in the world. As the name states, it is an adaptation of the ICD-9 codification system performed by the U.S. Health Department in collaboration with the Medicare and Medicaid Service Centres [6]. With more than 13,000 diagnostics and 3,500 procedures, it is essential to develop tools or systems focused on the codification process. Some Web applications like "Find-A-Code" or the work of Marisa Teresa Chiaravalloti et. all in "A Coding Support System for the ICD-9-CM Standard" are examples of systems developed for codification purposes [7]. The late one processes text in natural language using text mining algorithms, returning a list of possible codes for each case.

After the codification process, every discharge report can be read and perfectly understood in every country that adopts the same terminology. For the purpose of the present work and as stated by international directives, the ICD-9-CM codes are the basis of the DRGs decision [2].

C. AIDA and AIDA-PCE

Some Portuguese systems are equipped with a platform called AIDA *(Agência para Integração, Difusão e Arquivo de Informação Médica)* that uses intelligent agent systems that ensure the interoperability between different and heterogeneous information systems (SAM, SONHO, PCE, SAPE, RIS, LIS, among others) [8][9]. The electronic health record (EHR) is responsible for the safe and organized storage of all the information about a patient, from personal data to diagnostics and procedures [10]–[13]. The constant update is vital in this scenario, so the same Algoritmi group from Minho University that developed AIDA put together the AIDA-PCE. Following the Problem Oriented Medical Record (POMR), all the patient information like symptoms, medical observations, diagnostics and treatment plans are stored inside that structure.

Although using free text and other non-universal information the AIDA and AIDA-PCE present innovative and novel solutions to accomplish interoperability. On the other hand, the incorporation of the ICD-9-CM codes into the EHR represents an important feature to accomplish a cross-border medical record. The storage of ICD-9-CM coded discharge report saves space on AIDA-PCE databases and reduces medical errors in near future.

The next section presents the main research methodologies followed to implement this work.

III. Research Methodologies

The realization of any study in the field of Information Technologies (ITs) includes the scrutinized research and analysis of the set of methodologies and technologies available and feasible in the design of the defined IT solutions. The choice of the most appropriate methods and tools is mostly based on the advantages pointed out, as well as on associated limitations and compliance issues with related systems.

Thus, the achievement of the SPA for ICD-9-CM is based on the research methodology Design Science Research (DSR), mostly used in the construction and evaluation of useful and rigorous IT solutions. Each of the design phases presented in this study included the choice and use of the most appropriate methodologies, technologies and tools for the definition and elaboration of the desired solution. Finally, a Proof of Concept (PoC) was also carried out corroborating the viability and usefulness of the clinical coding tool for ICD-9-CM designed and developed, which consisted essentially of a SWOT analysis (Section V).

The next section presents the SPA for ICD-9-CM codification developed and implemented.

IV. Single-Page Application for ICD-9-CM Codification

The codification of the discharge reports was made manually in CHP, making the process too slow and with a high error probability. Thus, it emerged the need to create a process that would reduce the codification time. Therefore, a SPA was developed through which the health professionals are capable to perform the codification process, and at the same time to consult patients' data, such as the discharge report, the personal information and the hospital services were the patient was admitted.

The purposed layout in this scenario encompasses a solution with three main components: the patient information, the codification area and the discharge report. The codification area presents five frames: diagnosis, external causes, procedures, tumor morphology and observations. With exception of the observations board, which is the only one that allows free text insertion, all the other boards are composed by rows divided mainly in priority, description and code.

The main characteristic that makes this process faster than the already existing is the dynamic and aided search. When a word is typed in the description field a list of all the ICD-9-CM codes is presented, and when the user picks one of them on the description the respective code is automatically filled. On the other hand, when the user enters the code the description is also automatically filled.

When a codification is finished, the user sends it to an evaluator. If the codification fails in this evaluation phase for any reason, it is sent again to the list of codifications that needs to be done. Thus, the application was developed with three different modules. One to be used in the first codification, another to be used when a discharge report was already codified and for some reason failed in the evaluation phase, and one final module to be used when the user only has view permissions. In this last module, the users are not able to change any field present in the codification.

The application development leaned on the LAMP architecture. It uses Linux as the operative system, Apache as the Web server, MySQL as the relational database management system (RDBMS) and PHP as the object-oriented language.

In order to develop a fluid and dynamic application, the AngularJS framework greatly contributed. The modularity and extensibility of this JavaScript framework allows the development of diverse and futuristic applications.

The database storages all the data related to the codification codes, the discharge reports, user information, and all the information generated by the codification process. The RESTful Web service mediates the communication between the SPA and the database.

The next section presents the SWOT analysis of the Web application developed.

V. SWOT ANALYSIS

To test the viability, the utility, the quality and the efficiency of the application a PoC was necessary, in this case, a SWOT analysis. This analysis allows to analysis the strengths, weaknesses, opportunities and threats of the application [14].

Strengths:

- High usability, intuitive and easy to learn (user-friendly);
- Easy access to the data of patients, as well as the hospital services in which the patient was;
- High scalability;
- Easy of reissue of coded discharge reports;
- Decrease of the codification time of the discharge reports;
- Decrease of human error;
- Easy adaptability to different health institutions.

Weaknesses:

- Requires internet connection;
- Use of ICD-9-CM instead of the more current ICD-10-CM version.

Opportunities:

- Modernization and organizational development;
- Increasing expectation of the hospital administration to obtain methods that facilitate the hospital financing calculation;
- Provide the tool to help in the calculation of the hospital financing.

Threats:

- Lack of acceptance to resort to new technologies by health professionals.

The next section presents the conclusion and future work of this study.

VI. CONCLUSION AND FUTURE WORK

Finally, the realization of this case study allowed the development of a clinical practice tool, namely a user-friendly clinical coding tool for ICD-9-CM. The Web application is currently implemented in a production machine of CHP, and it is currently being used by the coding professionals of the hospital in order to perform the clinical coding of the episodes of hospital discharges from patients admitted to CHP. This will then facilitate the grouping of processes into DRGs, that is, a financial system that can manage the costs and waste associated with healthcare services. In the coming years, the expansion of the Web application is expected.

Trained professional using the ICD-9-CM clinical coding tool reported significant differences in time consumption and committed errors when using a computerized system to perform their tasks. It represents a valuable an asset to its users, since it facilitates the work of health professionals, and increases their capacity and speed of work by reducing the number of tasks required to perform a certain codification. In this way, the development of the clinical tool allows the centralization of a set of tasks and information in a SPA, greatly benefiting its users.

Regarding future work, the addition of a Business Intelligence (BI) module in the clinical coding tool for ICD-9-CM is foreseen, that is, the addition of a module with clinical and performance indicators [15]. Its principal aim is the visualization of indicators that show the association between the number of coded processes and each coding professional, as well as the temporal evolution of the number of processes encoded by each coding professional. Thus, the main objective of the insertion of this module is to study and analyze the performance of the coding professionals, that is, to identify, for example, the coding professionals who codify the most and also those who are coding the least. In this way, it is tried to encourage even more the increase of the production of the health professionals of CHP.

Finally, at the end of 2016, we already began the update of the system from the ICD-9-CM codification to the newer ICD-10-CM version.

ACKNOWLEDGEMENT

This work has been supported by Compete POCI-01-0145-FEDER-007043 and FCT - *Fundação para a Ciência e Tecnologia* within the Project Scope UID/CEC/00319/2013.

REFERENCES

[1] M. M. Bellanger, W. Quentin, and S. S. Tan, "Childbirth and Diagnosis Related Groups (DRGs): Patient Classification and Hospital Reimbursement in 11 European Countries," *Eur. J. Obstet. Gynecol. Reprod. Biol.*, vol. 168, no. 1, pp. 12–19, May 2013.

[2] W. Quentin, D. Scheller-Kreinsen, M. Blumel, A. Geissler, and R. Busse, "Hospital Payment Based On Diagnosis-Related Groups Differs In Europe And Holds Lessons For The United States," *Health Aff.*, vol. 32, no. 4, pp. 713–723, Apr. 2013.

[3] C. Fourie, N. Biller-Andorno, and V. Wild, "Systematically Evaluating the Impact of Diagnosis-related Groups (DRGs) on Healthcare Delivery: A Matrix of Ethical Implications," *Health Policy (New. York).*, vol. 115, no. 2–3, pp. 157–164, 2014.

[4] S. B. Rimler, B. D. Gale, and D. L. Reede, "Diagnosis-related Groups and Hospital Inpatient Federal Reimbursement," *RadioGraphics*, vol. 35, no. 6, pp. 1825–1834, Oct. 2015.

[5] Y. Xin, "Comparison of Hospital Medical Waste Generation Rate Based on Diagnosis-related Groups," *J. Clean. Prod.*, vol. 100, pp. 202–207, 2015.

[6] M. Rouse and D. Fluckinger, "ICD-9-CM (International Classification of Diseases, Ninth Revision, Clinical Modification)," 2014. [Online]. Available: http://searchhealthit.techtarget.com/definition/ICD-9-CM. [Accessed: 31-Jan-2017].

[7] M. T. Chiaravalloti, R. Guarasci, V. Lagani, E. Pasceri, and R. Trunfio, "A Coding Support System for the ICD-9-CM Standard," *Proc. - 2014 IEEE Int. Conf. Healthc. Informatics, ICHI 2014*, pp. 71–78, 2014.

[8] H. Peixoto, M. Santos, A. Abelha, and J. Machado, "Intelligence in Interoperability with AIDA," in *Foundations of Intelligent Systems*, Springer Berlin Heidelberg, 2012, pp. 264–273.

[9] J. Machado, A. Abelha, J. Neves, and M. Santos, "Ambient Intelligence in Medicine," in *IEEE Biomedical Circuits and Systems Conference*, 2006, pp. 94–97.

[10] M. Salazar *et al.*, "Step towards Paper Free Hospital through Electronic Health Record," in *Advances in Information Systems and Technologies*, Springer International Publishing, 2013, pp. 685–694.

[11] J. Duarte *et al.*, "Data Quality Evaluation of Electronic Health Records in the Hospital Admission Process," in *EEE/ACIS 9th International Conference on Computer and Information Science*, 2010, pp. 201–206.

[12] J. Duarte, C. F. Portela, A. Abelha, J. Machado, and M. F. Santos, "Electronic Health Record in Dermatology Service," in *ENTERprise Information Systems*, Springer Berlin Heidelberg, 2011, pp. 156–164.

[13] F. Portela *et al.*, "Electronic Health Records in the Emergency Room," in *IEEE/ACIS 9th International Conference on Computer and Information Science*, 2010, pp. 195–200.

[14] R. Pereira, M. Salazar, A. Abelha, and J. Machado, "SWOT Analysis of a Portuguese Electronic Health Record," in *Collaborative, Trusted and Privacy-Aware e/m-Services*, Springer Berlin Heidelberg, 2013, pp. 169–177.

[15] A. Brandão *et al.*, "A Benchmarking Analysis of Open-Source Business Intelligence Tools in Healthcare Environments," *Information*, vol. 7, no. 4, p. 57, Oct. 2016.

Pervasive Business Intelligence in Intensive Medicine

An overview of clinical solution

Ana Pereira, Filipe Portela*, Manuel Filipe Santos, José Machado, António Abelha and Fernando Rua
Algoritmi Research Centre, University of Minho, Guimarães, Portugal
Intensive Care Unit, Centro Hospitalar do Porto, Porto, Portugal
{cfp,mfs}@dsi.uminho.pt; {jmac,abelha}@di.uminho.pt

Abstract— **In the Intensive Care units, the information and data collected have a high value. It can help the physicians to take the best decision in the patient best interest. In this field, it is imperative to have the information available anywhere and anytime using Business Intelligence Concepts. Pervasive Business Intelligence (PBI) arises as a new technological field. This solution is focused on a person-centered approach, where the patients and the clinicians are the core. In this paper, the Pervasive Computing area is addressed having in consideration the Pervasive Health Care and the Intensive Care units. At first, the Strengths, Weaknesses, Opportunities and Threats analysis are presented, and then an overview of a PBI solution is described.**

Keywords-Pervasive; Business Intelligence; Intensive Care Unit

I. INTRODUCTION

The technological developments related to the diminishing of electronic components have been enormous. At the same time, the creation and spread of the Internet, connecting all machines and providing the worldwide sharing of information capacity reinforce the value of computing. Also the developing new techniques and new technologies, on par with an ever-growing capacity of miniaturisation of the devices and their cost reduction, developed the research capacity in this area.

Huge advances were achieved regarding the portability and storage capability, as well as new arrangements of human-computer interaction and wireless communication technologies, allowing to find computing and communication technologies anywhere and anytime. This situation has allowed and potentiated new interactions between people and their physical environment.

Especially in the field of healthcare, the Business Intelligence (BI) systems advancements have been enormous and especially in the latest years, many solutions have been implemented, given the vast quantity of data to gather and process. In the intensive health care units, the amount of data to pick increases exponentially, many times collected in real time. The need to make available the gathered data is of vital importance for the health professionals in these units. To access the patient's history, at any time and from any place starts to be a need for these intensivists.

In this article, we aim to understand the concept Pervasive, its appearance and joining of the Pervasive Healthcare with the BI systems in this field. This work may, therefore, serve as a guide to those who, in the future, aim to insert Pervasive in the BI systems.

This paper is divided into four sections. After a briefly Introduction, the related concepts are presented in Background Section. Then, in Section 3, the Strengths, Weaknesses, Opportunities and Threats (SWOT) analysis is presented, and a brief solution analysis is done. Finally, some conclusions are written, and the future work is mentioned.

II. BACKGROUND

This section presents the related concepts of the work.

A. *Pervasive HealthCare*

The definition of the term Pervasive is, according to the free dictionary, *"to spread something widely and deeply; affecting all aspects of something."* As in the Merriam-Webster dictionary, Pervasive means *"to exist in each part of something; spreading to all parts of something."*

Applying the concept Pervasive to the health field arises the Pervasive Healthcare (PHC) concept. PHC is considered an important area of research [1]. These systems have a dynamic structure and configuration, and an adequate understanding of these structures and the communication between its components, as well as the warranty of a proper and timely execution, is crucial. Pervasive Healthcare is one of the developing technologies using the Pervasive computing paradigm [2]. Pervasive computing presence provides an innovative way for the data transmission from medical applications. Currently, it is used wireless technology in various health domains [2].

For Rafe and Hajvali [1], Pervasive Healthcare is characterised by complex information, a dynamic number of interesting parts (stakeholders), and by ubiquitous computing, which connects perfectly the digital infrastructures in our daily lives. It gathers, processes and distributes "any kind" of personal information and contextual data at any time [1]. Pervasive Healthcare demands knowledge of the standard individual functions so that it provides the advanced detection of diseases, changes in functionality. It also provides pro-active prevention, as well as health services to predict individual well-being. This means that Pervasive Healthcare requires information that covers a person's entire life, including data on the personal behaviour, lifestyle, emotions, genealogical data, social data, psychological functionality and environmental sensors data [1].

Pervasive Healthcare is considered a key factor in the reduction of expenses and it is known for allowing improvements in disease management. The advances in communications technologies and wireless networks provided the acquisition, transmission, and treatment of critical medical information in real time [8].

B. *Understanding Ubiquitous Computing*

For M. Weiser [3], Pervasive computing is the most recent paradigm in computing, known as Ubiquitous Computing (Ubicom). Saha [4] foreseen that the Pervasive Computing (PerCom) devices would not just be personal computers. PerCom is also much smaller devices. It can be invisible appliances incorporated into almost any type of object that you can imagine, including cars, tools, household appliances, clothing and various consumer goods, all communicating through continuously more interconnected networks.

Currently, Ubiquitous computing represents a new direction in researchers about the integration and use of computers in people's lives and aims to reach a new computing paradigm, in which exists an elevated degree of penetration and ample availability of computers and other IT devices, with communication capabilities in our physical environment.

Beyond our personal computers, various computing devices incorporate physical places and interconnected objects fixed or mobile, the latter usually being wireless.

Ubiquitous computing already represents, in fact, a new computing paradigm, including the assumption that computers should "disappear" in the physical environment, becoming an integral part of such.

The Pervasive systems and ubiquitous technologies are ever more present in the domains of almost all businesses, improving the method of fulfilling them. In more personal or social areas, they are used principally to improve people's quality of life.

As examples of technology innovations, we have iPad Mini and the iPhone 6 (both devices are from Apple) or the Samsung Galaxy Tab. These devices have, for example, A-GPS, camera, microphone, with 7-9 hours of battery, 4G Internet and Wi-Fi connection, digital compass, accelerometer, environment light sensor, etc.

C. *Understanding Pervasive Computing*

For D. Saha [4], Pervasive computing (PerCom) would be, in the future, "omnipresent" combining ubiquitous open-source applications with the quotidian activities of the human being.

From PerCom's point of view, the environment would be saturated with a series of computing and communication resources very well blended with daily life. It allows the user to interact with a smart environment from everyplace, using an apparently invisible infrastructure from various devices/communication/computing, fixed and wireless networks. PerCom would create a digital omnipresent, sensible and adaptable environment for the human needs, characterised by the following essential elements: Pervasiveness (omnipresence), transparency (invisibility) and intelligence [4].

Pervasive computing would provide surprising enhancements in our capacity to connect and communicate [9]. It would gradually become integrated into our lives and daily activities, through natural ways of human-computer interaction, such as it has been verified. Even currently, the benefits and applications of Pervasive computing are far from being finished.

Various business fields, such as insurance companies, government agencies, health organisations, etc., can still get multiple benefits from Pervasive computing.

What was initially limited to the development of technology to make Pervasive computing more than a vision, as clearly gone beyond the first frontiers reaching the development of applications for various organisational domains [5]?

D. *Pervasive Mobile Architecture*

PerCom, as mentioned in [4], was pointing to the world where every object, every building, and even everybody would become part of a network service, where there is an expectation that the number of Pervasive devices would multiply rapidly.

Ideally, PerCom should approach all and every device on the globe with embedded active/ passive intelligence. These new smart gadgets or smart devices would be fit into microcomputers that would allow the users their connection to intelligent networks, and therefore, gain direct access, straightforward and secure, to information and services. These devices would then be mainly known as Pervasive devices [6].

One of the most common, current examples are the GPS based sensors, which provide location data that is translated into an internal representation of latitude, longitude, and altitude. Then advancing in time, we can verify that for Prabhakar et. al [6], there is more and more advances in mobile technology devices, such as, for example, smartphones, these technologies will become more Pervasive and omnipresent. Therefore, this verified reality in 2015, was in 2005 just a forecast/vision almost impossible to believe.

E. INTCare

The INTCare system was elaborated by Portela et al. [7] and implemented in the Intensive Cares Unit (ICU) of the Santo Antonio's Hospital, Porto's Hospital Center (CHP). It is an intelligent decision making support Pervasive system, composed of a group of integrated modules that execute all chores regarding knowledge discovery in an automated way and in real-time. In accordance to Portela et. al [7] INTCare can present anywhere and at any time information/knowledge, essential for the clinic decisions and whose primary purpose is the blending of a group of data sources to obtain interoperability advantages and the use of Data Mining models.

The INTCare System has improved the way the data is gathered and how it allowed the performance of manual actions made by the ICU professionals. It allows an entirely automatic collecting data process, in real-time, and besides that, it avoids paper registry [7].

In [8], it is referred that the INTCare health system consists of a Smart Decision Making Support System and Pervasive (SDMSSP), focused on the field of intensive medicine, developed in the unit of intensive care at Santo Antonio's Hospital, with the following characteristics: Online training; Real-time; Adaptability; Data Mining models; Decision Models; Smart Agents; Precision; Security; Pervasive/Ubiquity; Privacy; Safe access from the exterior and Use policy.

III. RESULTS

The achieved results are divided into two groups: SWOT and web application.

A. SWOT

Figure 1 presents a SWOT analysis of the Pervasive concept inserted in Business Intelligence Systems, developed while keeping in mind the context of critical healthcare area.

Figure 1. SWOT Analysis

B. Application

In this work, several dashboards were developed (sum, average, mode, max, and min). These panels address Admissions from emergency, urgency, other ICU, other hospitals, other, Admission type (urgent or programmed), disease, vital signs, ventilation, therapeutics (medications, dosages), patient information (sex, age), clinician, date (day, month, year, hour), type and others. In the next images, it is possible to observe some of the contents developed. Figure 2 refers to a dashboard where it is possible to verify the total number of patients admitted to the Intensive Care Unit (ICU). Here, we can know the admissions regarding the origin of the patients by service (SERVICO, DESP) and the number of admissions by sex (SEX), which, in this case, they were mostly male, with 1738 cases of this sex and only 1139 of the women. In Figure 2, it is also possible to see that 102 patients had a cardiac issue (insufcardiac) and 36 used a pacemaker; 40 patients did chemotherapy and 28 a radiotherapy. Finally, 123 patients were taking insulin.

Figure 2. An example of Dashboards developed - total number of patients admitted to the Intensive Care Unit.

In Figure 3, it is possible consulting the data collected from patients admitted to the ICU of CHP and some of their characteristics, such as the episode number of hospitalisation (EPISODIO), process number (PROCESSO), sex (SEX), date Birth (Year, Quarter, Month, Day).

During the period of October 2010 - March 2016, ICU had 2877 patients (1738 Males and 1135 females) admission with an average of 66,91 years old. 50% of the admissions came from either ICU or cardio service (UIHC).

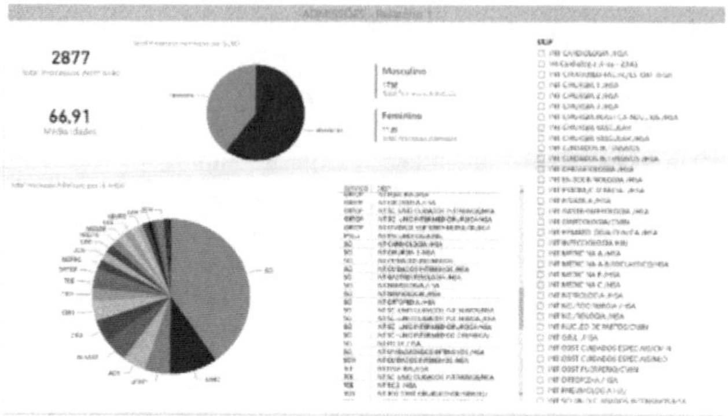

Figure 3. An example of Dashboards developed - data from patients admitted to the ICU of CHP.

IV. CONCLUSION

This paper presents the first overview of a Business Intelligence solution developed in an Intensive Care Unit. In this study, the SWOT analysis shows that the Information Security is the primary concern and the data availability (anywhere and anytime) is the main benefit of PHC. Complementary, this paper also showed the first application images.

After a first analysis, the intensivists consider this application interesting; however, it needs a deeper clinical analysis done by them. A Business Intelligence Application provides only the data in an intuitive way. Then the professionals make the results analysis with the focus on the contribution of it to their decision-making process.

In the future, the intensivists will perform a deep analysis of the dashboards. Other panels are being developed, and then the final solution will be deployed in the real environment.

ACKNOWLEDGEs

This work has been supported by Compete: POCI-01-0145-FEDER-007043 and FCT - Fundação para a Ciência e Tecnologia within the Project Scope UID/CEC/00319/2013.

REFERENCES

[1] V. Rafe and M. Hajvali, "Designing an architectural style for Pervasive Healthcare systems.," *J. Med. Syst.*, vol. 37, no. 2, p. 9927, Apr. 2013.

[2] G. Sudha and R. Ganesan, "Secure transmission medical data for pervasive healthcare system using android," in *2013 International Conference on Communication and Signal Processing*, 2013, pp. 433–436.

[3] M. Weiser, "The computer for the 21 st century," *ACM SIGMOBILE Mob. Comput. Commun. Rev.*, vol. 3, no. 3, pp. 3–11, Jul. 1999.

[4] D. Saha, "Pervasive Computing: A Vision to Realize," in *Advances in Computers*, vol. Volume 64, V. Z. Marvin, Ed. Elsevier, 2005, pp. 195–245.

[5] [K. Lyytinen and Y. Yoo, "Issues and Challenges in Ubiquitous Computing," *Commun. ACM*, vol. 45, no. 12, pp. 63–65, 2002.

[6] T. V. Prabhakar, U. Mysore, U. Saini, K. J. Vinoy, and B. Amruthur, "NFC for Pervasive Healthcare Monitoring," in *2015 28th International Conference on VLSI Design*, 2015, vol. 2015-Febru, no. February, pp. 75–80.

[7] F. Portela et al"Pervasive and intelligent decision support in Intensive Medicine – the complete picture." Springer, 01-Jan-2014.

[8] [F. Portela, M. F. Santos, and M. Vilas-Boas, A pervasive approach to a real-time intelligent decision support system in intensive medicine. Springer, 2012.

Future Directions of Applying Healthcare Cloud for Home-based Chronic Disease Care

Yan Hu, Sara Eriksén
Department of Creative Technologies
Blekinge Institute of Technology
Karlskrona, Sweden
e-mail:{yan.hu, sara.eriksen}@bth.se

Jenny Lundberg
Department of Media Technology
Linnaeus University
Vaxjo, Sweden
e-mail: jenny.lundberg@lnu.se

***Abstract*— The care of chronic disease has become the main challenge for healthcare institutions around the world. To meet the growing needs of patients, moving the front desk of healthcare from hospital to home is essential. Recently, cloud computing has been applied to healthcare domain; however, adapting to and using this technology effectively for home-based care is still in its initial phase. We have proposed a conceptual hybrid cloud model for home-based chronic disease care, and have evaluated its future feasibility by a case study of diabetes care in Blekinge, Sweden. In this paper, we discuss some possible future opportunities and challenges to apply this cloud model with the huge population for home-based chronic diseases care. To apply this model in practice, a professional IT healthcare education team is needed for both healthcare providers and healthcare recipients. For home-based healthcare, a monitoring system with an automatic alarm to healthcare providers is also necessary in some cases. Also, how to record and integrate excises data through wearable devices in a cloud should be considered. Given the high demand, sharing medical images through the cloud should be another research focus.**

Keywords-future trends; hybrid cloud; chronic diseases; home-based care

I. INTRODUCTION

The care of chronic disease has become the main challenge for healthcare institutions around the world. According to a WHO report in 2014, almost 90% of deaths in Sweden are caused by chronic diseases [1]. As the incidence and prevalence of chronic diseases continue to increase, traditional hospital-based healthcare is less able to meet the requirements of patients. To meet the growing needs of patients, moving the front desk of healthcare from hospital to home is essential. Home-based healthcare could enable the care recipients to live independently at home. Healthcare providers could reach the patients based on their shared daily health data, and provide clinical suggestions. Also, for home-based healthcare, more people are encouraged to assist with the care, such as family members and other patients with similar symptoms.

The development of Information Communication Technology (ICT) has enabled people to enter a modern digital society. Our quality of life is promoted by the application of ICT in all fields. In the healthcare domain, by widely using personal computers, smartphones, and other self-monitoring devices, ICT has brought healthcare in a new era [2]. All of the ICT technologies could help improve the quality of home-based healthcare.

In recent years, cloud computing has been used to support healthcare. The obvious advantages of cloud computing, like big data storage, pay-as-go online software services, and powerful data analysis capacity make it a great benefit to apply cloud computing in home-based healthcare [3]. In our previous work, we have proposed a conceptual hybrid healthcare cloud model for home-based care. Moreover, we have tested the feasibility of the model by a case study of diabetes care in Blekinge, Sweden [4]. We believe that cloud-based solution is a future trend of healthcare technology. To prefect and realize our hybrid healthcare cloud in large-scale population, there are other technologies to be further explored.

This paper mainly discusses future opportunities and challenges of applying cloud model with the huge population for home-based chronic diseases care. Section 2 is a brief introduction of our research project, while in Section 3, we give a brief description on our Swedish case study. Section 4 lists some opportunities and challenges are discussed in Section 5. Conclusion and the suggestions of future work conclude the paper.

II. HEALTH IN HAND PROJECT

The Indo-Swedish R&D project Health in Hand – Transforming Healthcare Delivery is a project funded by VINNOVA (Reg.No.2013-04660) for three years. The main objective of the project is to establish long-term Indo-Swedish R&D collaboration around leading-edge applied health technology, with a focus on mobile health services, namely mHealth. mHealth technologies usually mean "medical and public health practice supported by mobile devices, such as mobile phones, patient monitoring devices, personal digital assistants (PDAs), and other wireless devices [5]". In this project, we mainly refer to "mobile services which inform, motivate and enable individuals to manage their own health information and knowledge sharing, as well as support communication and community building among both patient and caregiver communities [6]". The project uses participatory design as a research approach, to focus on the local design of design methods, techniques, and tools that promote participatory design for mHealth development [7].

One of the aims of the project is to discover possible solutions to enhance collaboration around the complex systems which are the base for innovative mobile technologies in healthcare among different healthcare providers and between healthcare providers and recipients. In the final year of the project, the study is focusing on concerning up-scaling and commercialization of mHealth technologies in collaboration with industry and public sector [6]. In the beginning of the project, we tried to develop applications based on mHealth technologies. When we had our first Swedish case study, we found that there were already a number of mobile applications in the market, so we move our research focus to how to integrate these mobile applications into current healthcare system.

III. A CASE STUDY IN SWEDISH DIABETES CARE

In our previous work, we have studied on how to adapt cloud computing to healthcare domain and have proposed a conceptual hybrid cloud model for home-based chronic diseases. In this conceptual model, healthcare recipients using mHealth technologies for self-recording home-based healthcare data, and communicating with their healthcare providers, is one of the key elements. Within the scope of the project, we have conducted a two-step case study in diabetes care. The purpose of this case study was to improve the conceptual model and test its feasibility on one hand, and to explore some possible technical trends for up-scaling and commercialization of this model on the other hand [4]. The improved model is shown in Figure 1.

This model is a hybrid cloud model, which contains private clouds from the hospital and primary care centers, as well as public clouds for home-based self-management data. This model is based on that diabetes patients can measure their blood glucose and other parameters themselves at home via a few devices and upload the values to the app's public cloud automatically. The secure connected public cloud of the app not only stores the data, but also deals with the data with the help of powerful data processing capacity of cloud computing to monitor the threshold values and create different kinds of reports. The doctors at the hospital can access the public cloud if they have the authorization of their patients. Between the hospital's and primary care centers' private clouds, there is an internal journal system for sharing partly patients'' data sharing. All these ways of sharing data already existed in Blekinge, as well as other counties in Sweden. The improved communication between patients and primary care providers is the most pressing demand of the day.

This model visualizes the current data sharing of home-based diabetes care, as well as provide a cloud computing solution to enhance collaboration around the healthcare systems and moblie applications based on mobile technologies. For other types of chronic diseases, this model could also be used for data sharing. From our interviews with both healthcare providers and healthcare recipients, we found that it is feasible to apply this hybrid cloud in future home-based healthcare. For the model's up-scaling and commercialization, there remains future work to be further explored.

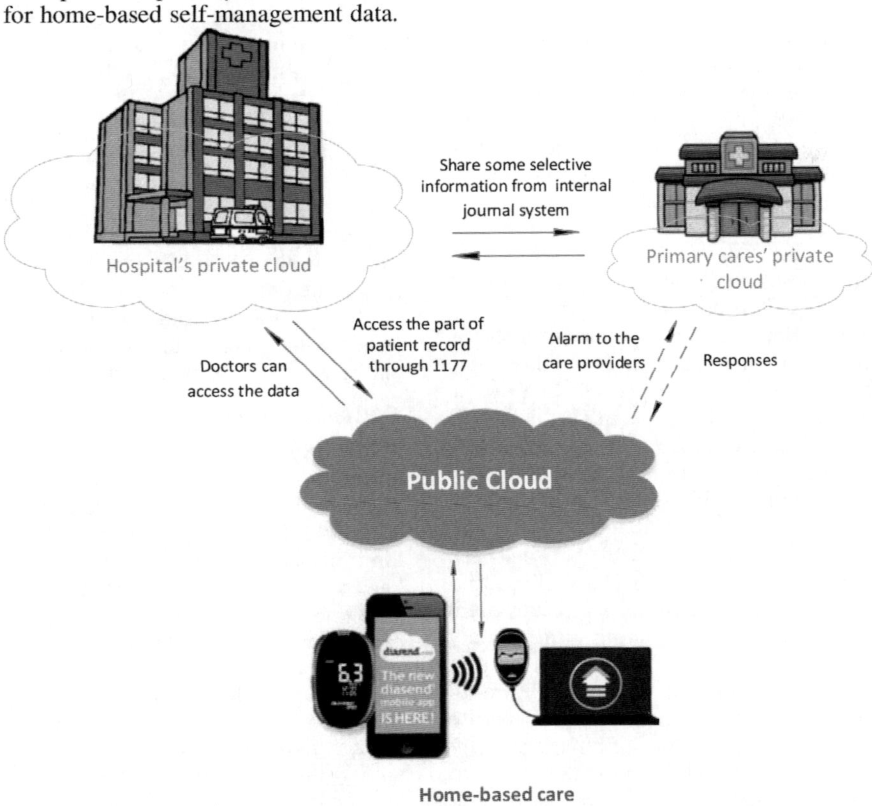

Figure 1. Conceptual hybrid cloud model [4]

IV. Future opportunities

A. Professional healthcare IT education

From our case study, we found that all eleven participants feel ICT technologies have developed very fast, and it is quite difficult for them to follow the steps of the development. The lack of IT knowledge sometimes reduces the enthusiasm of users to try new ICT technologies in healthcare. Especially with mobile technologies used for home-based healthcare, healthcare recipients thought it was healthcare providers' responsibility to teach them to use the mobile applications or self-recording devices. However, most of the healthcare providers thought they were not good at technology themselves. Although there are use guides from the service or device providers, the healthcare recipients still ask lots of technically related questions to their healthcare providers. From our case study, we were surprised find that healthcare providers are less interested in adopting new ICT technologies than healthcare recipients. The reason is that they felt their main work is diagnosis and treatment, and that using new technologies reduced their work efficiency in some way.

In this situation, building up a professional healthcare IT education team is of utmost importance. This education team should be built by people who have certificated knowledege in healthcare technology. It could be supported by government or set up as an independent commercial organization. The educational programs provided by this team should address both healthcare recipients and healthcare providers. For healthcare providers, training on their working systems and the communications between these systems could be offered frequently, like once a year. Since in chronic diseases care, self-management is a key factor [8], the education to help healthcare recipients to use different ICT technologies for self-management would be the main focus.

B. Mobile doctor

Another trend of home-based chronic diseases healthcare is the mobile doctor. The mobile doctor here refers to having a communication with a primary care doctor through a phone, tablet or computer. Healthcare recipients could receive primary care at any place of their convenience. In Sweden, there are two of this kind of services on the market, KRY [9] and Min Doktor [10]. The difference between these two services is that KRY provides a booked 15 minutes video meeting with doctors while Min Doktor provides the communication with the doctors with messages all day round. Both these two services are linked with personal ID; healthcare recipients must log in with their social security number. This will guarantee the healthcare recipients could receive the same benefits as they are visiting local primary healthcare centers.

Both these services are still in their beginning phases. In the future, we believe more and more chronic disease healthcare recipients will choose this type of remote primary care to communicate with their healthcare providers. How to promote and improve the quality of these mobile doctor services should be under discussion.

C. Real-time monitoring system

In Sweden, most patients with chronic diseases now contact more often with primary a healthcare center than with a hospital, which means primary healthcare centers take more responsibility for chronic diseases care. However, from our case study, we found that the communication between patients and primary healthcare centers is still based on regular mail and telephones.

As big data in the cloud brought a revolution in healthcare [11], a real-time monitoring system with alarm could be a possible mechanism for primary care. The calculated threshold values by the big data in the public cloud, together with healthcare providers in primary healthcare centers, can help the patients to set threshold values for their daily health parameters. When a patient's self-recording data is uploaded at home, and the values reach the threshold values, the data processing program in the public cloud can send an alarm to the responsible primary care center. As soon as the care providers see the alarms, they could contact the patients and give them some feedback. The design and development of this real-time monitoring system could be another future direction to apply a healthcare cloud for home-based chronic diseases care.

D. Wearable devices to record excise data

According to WHO, lack of physical activities is a significant risk factor for chronic diseases such as stroke, diabetes, and cancer. However, an estimated 23% of adults and 81% school-going adolescents are not active enough [12]. In most cases of chronic disease care, healthcare providers suggest their patients do daily physical excise as the main method to control their condition. With the widely use of wearable devices, it is possible for people to record their excise data[13]. Today most data recorded by the wearable devices is stored in device companies' servers. With the 5G network [14] and several sensors that are being introduced into the healthcare environment, and which provide even more healthcare data.

In the future, integrating this excise data with other home-based self-management data from both mobile applications and sensors will be an important trend.

E. Medical images sharing

The huge data storage capacity in the cloud makes it easier to store and process medical images. This will promote medical image sharing between healthcare providers and recipients. In Blekinge, Sweden, until now, there is no such kind of medical image sharing system, but the demands are keen according to our case study, especially for aged people who live independently. They want to bs able to take photos of wounds at home and send them to the healthcare providers in wound centers, which would be better than regular visiting. Moreover, the healthcare providers in wound centers also thought it could save time to treat the wounds.

A multi-functional medical image sharing mobile application could be a future research orientation. For example, how to pre-process the wound images before the healthcare providers review them, such as wound highlight, 3D visualization, are worthy of consideration.

F. Games for health

Mobile games have a significant influence on people's physical activities, as illustrated by the success of Pokémon Go [15]. Games for health are not limited to serious games anymore. Designing an entertaining and fun digital game to encourage patients with chronic diseases to do more excises or an educational mobile game to help the healthcare recipients to get more knowledge of their diseases could be a new interest for both game and healthcare industry.

V. FUTURE CHALLENGES

Challenges to future development of healthcare cloud in home-based chronic diseases care also need to be carefully considered and addressed

A. User-friendly information system design

Most patients with chronic diseases are aged people. Thus the new information systems designed for home-based chronic diseases need to be user-friendly. In additional, smart tablets are easier to use for aged people than smartphones since the screen is bigger.

B. Data leakage of public cloud

In home-based healthcare, most self-recording data from home environment is stored in public cloud. Thus, it is a risk of data leakage. Access control and data security protection are always necessary for healthcare related data.

C. Reliability of third party healthcare services

There are a huge amount of healthcare mobile applications to support home-based chronic disease care, but not all of them are reliable. According to study on medical related mobile apps, less than half of the studied apps are reliable[16], which means if healthcare recipients use these unreliable apps, they may get wrong information and advice about their diseases. To overcome this challenge, the companies provided healthcare services, responsible government agencies and mobile app users should make efforts together.

VI. CONCLUSION AND FUTURE WORK

We found that our cloud model is feasible to be applied in home-based chronic disease care in the future. It will help to promote patient-centric healthcare in Sweden. In this paper, we mainly address the future trends of applying healthcare cloud for home-based chronic disease care. To upscale and commercialize the healthcare cloud, professional healthcare IT education, mobile doctors, real-time monitoring systems, wearable devices to record excise data, medical image sharing and games for health could be the main future directions. User-friendly design, data privacy and security, and reliability of services would be challenges to promote healthcare cloud development. In the future, we will put the focus on one or two above issues in our continued research.

ACKNOWLEDGMENT

This work was partly supported by VINNOVA project" Health in Hand – Transforming Healthcare Delivery" (Reg. No. 2013-04660).

REFERENCES

[1] "WHO | Noncommunicable diseases country profiles 2014," *WHO*. [Online]. Available: http://www.who.int/nmh/publications/ncd-profiles-2014/en/. [Accessed: 02-Feb-2017].

[2] S. S. Deo, D. N. Deobagkar, and D. Deobagkar, "Design and development of a web-based application for diabetes patient data management," *J. Innov. Health Inform.*, vol. 13, no. 1, pp. 35–41, 2005.

[3] Y. Hu and G. Bai, "A cloud model for interoperable Home-based Chronic Diseases Healthcare," in *2014 World Symposium on Computer Applications Research (WSCAR)*, 2014, pp. 1–6.

[4] Y. Hu, G. Bai, S. Eriksén, and J. Lundberg, "Hybrid cloud model for diabetes some-based care -a case study for perceived feasibility in the future." unpublished.

[5] "mHealth New horizons for health through mobile technologies." [Online]. Available: http://www.who.int/goe/publications/goe_mhealth_web.pdf. [Accessed: 03-Feb-2017].

[6] S. Eriksén, M. Georgsson, M. Hofflander, L. Nilsson, and J. Lundberg, "Health in hand: Putting mHealth design in context," in *2014 IEEE 2nd International Workshop on Usability and Accessibility Focused Requirements Engineering (UsARE)*, 2014, pp. 36–39.

[7] S. Eriksén, J. Lundberg, M. Georgsson, L. Nilsson, M. Hofflander, and C. Borg, "Transforming Healthcare Delivery : ICT Design for Self-Care of Type 2 Diabetes," in Nordic Conference on Human-Computer Interaction (NordiCHI), in press.

[8] H. Holman and K. Lorig, "Patient self-management: a key to effectiveness and efficiency in care of chronic disease," *Public Health Rep. Wash. DC 1974*, vol. 119, no. 3, pp. 239–243, Jun. 2004.

[9] "KRY - meet doctors online," *KRY - meet doctors online*. [Online]. Available: https://kry.se. [Accessed: 04-Feb-2017].

[10] "My Doctor: Meet doctors safely & easily in the app" [Online]. Available: https://mindoktor.se/. [Accessed: 04-Feb-2017].

[11] B. Kayyali, D. Knott, and S. V. Kuiken, "The big-data revolution in US health care: Accelerating value and innovation | McKinsey & Company." [Online]. Available: http://www.mckinsey.com/industries/healthcare-systems-and-services/our-insights/the-big-data-revolution-in-us-health-care. [Accessed: 04-Feb-2017].

[12] "WHO | 10 facts on physical activity." [Online]. Available: http://www.who.int/features/factfiles/physical_activity/en/. [Accessed: 02-Feb-2017].

[13] D. Son, J.Lee, S.Qiao, R. Ghaffari, J.Kim, J.Lee *et al.*, "Multifunctional wearable devices for diagnosis and therapy of movement disorders," *Nat. Nanotechnol.*, vol. 9, no. 5, pp. 397–404, May 2014.

[14] A. Osseiran, F. Boccardi, V. Braun, K. Kusume, P. Marsch, M. Maternia et al., "Scenarios for 5G mobile and wireless communications: the vision of the METIS project," IEEE Communications Magazine, vol. 52, no. 5, pp. 26–35, May 2014.

[15] T. Althoff, R. W. White, and E. Horvitz, "Influence of Pokémon Go on Physical Activity: Study and Implications," *J. Med. Internet Res.*, vol. 18, no. 12, Dec. 2016.

[16] F. Haffey, R. R. W. Brady, and S. Maxwell, "A Comparison of the Reliability of Smartphone Apps for Opioid Conversion," *Drug Saf.*, vol. 36, no. 2, pp. 111–117, Feb. 2013.

Technology in Health Care

A new research and teaching subject in collaboration between nursing science and engineering

Britt Östlund
Department of Technology and Health
Royal Institute of Technology, KTH &
Swedish Red Cross University College, SRCUC
Stockholm, Sweden
e-mail: britt.ostlund@sth.kth.se

Gunilla Björling, Janet Mattsson, Sara Stridh, Madeleine Sahlström
Swedish Red Cross University College, SRCUC
Stockholm, Sweden

Abstract— **Today, health care systems face a number of challenges related to technological developments. This work in progress is a new Swedish initiative for collaboration between nursing science and engineering focusing digitization, demographics and participation. The initiative aims at understanding how digitization affects patients and health care professionals and the role of engineers and how this collaboration proactively contributes to systems that support caring and nursing. The presentation expects input on the programs substance and boundaries and whether this initiative is fruitful to create sustainable health care systems.**

Keywords; collaboration, nursing science, caring science, engineering, patient support, digitization.

I . INTRODUCTION

Today, health care systems face a number of challenges related to technological developments. The Global Commission on Education of Health Professionals for the 21st Century describes a mismatch between the care offered and people's demands and needs; lack of cooperation; discontinuous care chains; tenacious hierarchies, and not least, a focus on technology founded on flawed understanding of the context in which the technology is used [1]. The gender system that locks structures what perceived as male and female work is deemed particularly difficult to change. Significant is also the lack of good examples of how to meet these types of challenges. Other publications confirm these results, reporting on a lack of accuracy in technological support for health care, not least the confidence in individual solutions. There are reports on an imbalance between the success factors in health care - increased life expectancy and better treatments - and the expectations of what you want done and what to expect [2] [3]. At the same time, the belief in technology as the ultimate solution still prevails. The Swedish Society for Nursing identifies a lack of knowledge about the impact of technology on caring interventions. To meet this, they emphasize individual health as an important aspect to investigate and call for possibilities to actively participate in design, implementation and assessment of new technology [4] [5]. Others stress the economic values of returning to patient centred strategies [6]. To mention one example, creating digital health care records and digital information to citizens are among the most difficult tasks to accomplish despite the wide access to Internet and digital infrastructures. Google declared a few years ago that their investment in Google Health was one of their biggest failures due to its complexity [7]. Instead they are now developing databases focusing on health and aging sciences. Another way to meet these challenges is through multidisciplinary initiatives between scientific disciplines and between academy and industry. Today there are a number of examples of international multidisciplinary collaboration, e.g., the Bio design Fellowship program at Stanford University since 2003 which is also implemented in Sweden as The Centre for Technology in Medicine and Health (CTMH). These types of commitments are key elements in promoting interdisciplinary research and development in medicine focusing problem-solving in general. However, it is rarer to find a technology focus in relation to nursing and caring.

Section 2 describes the rationale behind the organization of a new collaboration between engineers and nurses to meet the challenges pointed out above. Section 3 describes the aim more in specific with emphasis on digitization effects on inter-professional collaboration and the dependency on active patients to make technology work in their social context. In section 4 and 5 some arguments are presented to why nursing and critical technology research is a relevant combination to set the agenda for this new collaboration. Section 6 summarizes the main conclusions.

II. NEW INITIATIVE TO MEET THE CHALLENGES

With this background we would like to present a new Swedish initiative for collaboration between nursing science and engineering, initiated jointly between KTH, the Royal Institute of Technology and SRCUC, the Swedish Red Cross University College in Stockholm. The reason for submitting this work in progress at this conference is to gather input and comments on two main issues. The first one is whether this is a fruitful strategy to create more sustainable health care systems. The second issue is how we in a multidisciplinary collaboration can work to define “Technology in health care”, its substance and its boundaries.

The collaboration between KTH and SRCUC is implemented in parallel with a new interdisciplinary postgraduate program - Technology in Health Care (in Swedish: Teknisk vårdvetenskap) at KTH and as a part of the nursing science program curriculum at SRCUC. The Nursing Science program at SRCUC will therefore have a unique technological profile, added to its 150 year tradition of training nurses. At KTH, the Department for Technology and health will be given new opportunities for research into how technology works in caring settings.

The aim of the collaboration is to create sustainable and effective systems of health care through increased knowledge about how technology affects people's opportunities to regain their health and more proactive nurses and engineers participating in design and innovation processes which also enhance patient safety. The technology focus is digitization encompassing both biomedical engineering used by health care professionals in hospitals, and home health care systems, as well as digital tools used by both health care professionals, citizens at home and in mobile settings. Demographic developments will be key focusing on children and older populations, the groups that will increase the most until 2050. Many projects fall outside the boundaries of eHealth, yet include the combination of technical skills and caring skills to solve the problem at stake. Digitization is central. It is estimated that within a decade, the majority of the world’s population will have accesses to virtually all the world's information in a machine that fits in its own palm. These profound changes occurring within one generation naturally has a great impact on the Red Cross, which together with the Red Crescent, is the world's largest non-profit organization (97 million members). Digital media, robots and sensors, create new opportunities to practice disaster response and care, in dangerous situations and in the monitoring of health.

The SRCUC-KTH collaboration was initiated jointly in 2012, a professor was installed 2014 and 2016 we are moving from preparations and discussions to research and education activities supported by structures for training, qualification, publication and projects.

III. PROGRAM CONTENT AND GOALS

The aim is to understand how digitization affects patients and healthcare professionals and proactively contribute to functioning systems that support caring and nursing. With digitization we refer to the on-going development from a hospital- and function-based organization focused on patients, diseases, wards and elimination of risks with single technical applications, towards a focus on health, home health care such as cancer and palliative care, monitoring and communication on distance, digitized and accessible patient care records requiring active patients and citizens and more of inter-professional collaboration and teamwork between health care professionals. These shifts in the way care and contacts with citizens are understood is underway. Our analysis goes even further, focusing on-going developments and implementations of patient safety outside hospitals and between hospitals and homes, distance surgery and design of self-care devices, the use of robotics, genetically guidance and other new roles for nurses. The critical input from this program to influence this development is a proactive approach including the ability to collaborate with engineers, initiate and formulate relevant research questions. The theoretical foundation for this critical input will be Science-Technology-Studies research providing insights on the configuration of users and contexts and from Informatics and Implementation science providing knowledge on participation in implementation of health information systems. This is expected to offset the dominant paradigm of deterministic and paternalistic views leading up to a mismatch between the care offered and people's demands and needs; lack of cooperation; discontinuous care chains; tenacious hierarchies, and not least, a focus on technology founded on flawed understanding of the context in which the technology is used. The research environment at KTH and SRCUC in which this program is developed include other programs which will be expected to make considerable input to the multi-disciplinary approaches, such as care logistics engineering, ergonomics, monitoring, nursing and public health.

To be proactive requires knowledge about technological developments and the ability to collaborate with engineers and participate in design and innovation processes both for healthcare professionals and concerned citizens. The first attempts to joint learning opportunities for engineers and nurses were successful and will continue this year The exercise is organized as a part of the curriculum in which students in nursing and students to become medical engineers collaborating around a report from the Swedish Accident Investigation Authority about a serious failure at a Swedish hospital involving medical technology and a patient. The meeting received very positive feedback and created a broader perspective on each other's responsibilities and the role of technology in a real life setting. For the moment the research group includes one professor from the engineering department with a STS-background, two PhD in medicine and nursing, one PhD in medical education related to informatics, three assistant professors, two PhD students and two project employed persons. The first six years of this collaboration is based on joint funding from KTH and SRCUC and from external funding. For the moment the external funding include a China-Sweden collaboration building home health care and housing for ageing populations; development of medical devices and patient safety funded by Swedish and Norwegian foundations; child intensive care developments funded by Karolinska hospital; digital ethics funded by Swedish foundations; and Age Management funded by Swedish Industry and a Swedish Research Council. The professor leading this group has for thirty years been dedicated to research and development in aging, technology and design and was responsible for ten years for the Ageing and design program at Lund University, where she still holds a visiting professorship. Her publications include a range of projects with old people participating in design and implementation. The first articles in this context were published by Mattsson & Stevens [15] [16] and Björling on coated endotracheal tubes and central venous catheters with focus on patient safety [17] [18]. During the same period Östlund published several articles on robotics and ethics [19] [20] [21].

The result of this program are expected to reinforce the participation of citizens and health care professionals in designing future health care, and as a consequence broadening the innovation capability in health care and citizens' trust and confidence in the health care system. The most tangible results in the short term is expected to be technology that does not impair care but that facilitates and enhances people's opportunities to recover and maintain their health.

IV. THE RELEVANCE OF NURSING AND CARING SCIENCES

Nursing is central in the development of future health care. In the Nordic countries, nursing as a research subject, have for thirty years, developed in parallel with “caring sciences”. Both have been expansive including caring informatics and caring theory. Nursing, which is the broad international field, has its focus on guiding nurses in practice such as routines and regulations and patient safety. Caring sciences, published in Journals, such as the Scandinavian JN for Caring Sciences among other JN, originate from phenomenology and the interest to understand principles for utilization [8] [9] [10]. Caring science is today related to person-centred care, self-sufficiency and independence. Especially since the core of caring has been revealed as central, holistic, individualized and at the same time providing expert physical care combined with fulfilling emotional needs in an adaptive environment [11]. As digitization increasingly moves in to the realms of health and self-care, the relationship between the caregiver and the individual citizen, patient or care receiver, becomes more important. Result from innovation and implementation research have not yet been applied in this field but can be important to promote”self-care management, personalized medicine” and consequences of the demographic development [12] [13].

V. TECHNOLOGY DEVELOPMENT

Technology development today is leading to greater complexity. We are now entering a new phase where it is more about interconnected systems and no longer just individual applications. Today, the use of different types of IT applications is not unknown to anyone in health care. “APPS”, digital patient records, alarms, sensors for monitoring health, social robots, “robocats”, digital incontinence, remote surgery, decision support for diagnosis, balance training for stroke sufferers are all examples of products that can add value”. Many of these examples have been shown to increase the quality of care and have already become successful business solutions. At the same time, this raises awareness about the fact that individual technical artefacts are hardly the solutions to the health care problems. This leads to a number of questions which need to be addressed: How can we permanently and sustainably integrate new applications in health care? What is the best way to implement accurate solutions in health care with a comprehensive and ongoing digitization?

Another important question concerns what is called the technological imperative in relation to caring values: are we always obliged to do what can be done in terms of technological development? Or can we find ways to criticize such deterministic views? Although technology is closely associated with the development of modern medical care, the relationship to technological development is divided [14]. Here is a criticism that high-powered specialization risks creating problems and become counterproductive. The German philosopher George Henry Gadamer asserted, for example, that it is precisely in highly developed technical civilizations that the phrases "quality of life" and "whole" are expressed, because something has been lost (15). Meanwhile, with a critical perspective on technology development, we can see that technology has a special attraction and that it has become a force for change. It also becomes apparent that a narrow technical perspective sometimes tends to give healthcare professionals the role of managing technology instead of people. It also contributes to the technological imperative i.e., what is possible to be measured must be measured even if the benefit is unclear.

VI. CONCLUSIONS

This short paper suggests that a program based on multidisciplinary collaboration between nursing and engineering is one of the key strategies to make digitization in health care sustainable. Although this can be expected to solve many of the problems addressed in the literature it is not enough. A theoretical reflection is also needed that questions the prevailing paradigm where technology is regarded as a tool independent of the influence of the context in which it is used. Such critical input will open up for more proactive strategies including the understanding of how digitization affects patients and how to increase participation in design and implementation of health information systems.

REFERENCES

[1] J. Frenk et al. "Health professionals for a new century: transforming education to strengthen health systems in an interdependent world," The Lancet vol. 376, no. 9756, pp. 1923-1958, 2010.

[2] I. R. Hallberg, "Moving nursing research forward towards a stronger impact on health care practice," International Journal of Nursing Studies, vol. 46, pp. 407-412, 2009.

[3] Swedish Society of Medicine, Onödiga eller skadliga åtgärder i svensk allmänmedicin – ett diskussionsunderlag, [Unnecessary or harmful actions in Swedish general medicine - a basis for discussion], Swedish Society of Medicine, Stockholm 2013.

[4] Swedish Society for Nursing:

- Kunskapsluckor inom omvårdnad. En första inventering [Gaps within caring. A first inventory], Stockholm 2013.
- Strategi för sjuksköterskors arbete med eHälsa [Strategy for nursing and eHealth], Stockholm 2012.
- Värdegrund för omvårdnad [Valuebased caring], Stockholm 2010.

[5]S. Bakken, "Informatics for patient safety: a nursing research perspective," Annu. Rev. Nurs. Res, vol. 24: pp. 219–254, 2006.

[6] M. E. Porter and E. Olmsted Teisberg, "Redefining Health Care. Creating Value-Based Competition on Results," Harvard Business Review Press, Boston Mass. pp. 98-148, 2006.

[7] R. Kurzweil, The Singularity is Near. New York: Viking, 2005.

[8] V. Sorlie, L. Jansson and A. Norberg, "The meaning of being in ethically difficult care situations in paediatric care as narrated by female Registered Nurses," Nordic College of Caring Sciences, vol. 17, pp. 285-292, 2003.

[9] L. Fagerstrom, K. Eriksson and I. B. Engberg, "The patient's perceived caring needs: measuring the unmeasurable," Int. JN. Nurs. Pract, vol. 5, no. 4, pp. 199-208, 1999.

[10] P. E. Benner, C. A. Tanner and C. A. Chesla, Expertise in nursing practice : caring, clinical judgment & ethics, 2rd ed., New York: Springer Pub. 2009.

[11] J. Mattsson, M. Forsner, M. Castrén and M. Arman, "Caring for children in pediatric intensive care units: An observation study focusing on nurses' concerns," Journal of Nursing Ethics, vol. 20, no. 5, pp. 528-538, 2013.

[12] L. Damschroeder et al. "Fostering implementation of health services research findings into practice: a consolidated framework for advancing implementation science," Implementation Science, vol. 4, no. 50, pp.1-15, 2009.

[13] I. R. Hallberg, "Omvårdnadsvetenskapen i ett framtidsperspektiv [Caring science in future perspectives]," In E. Hamrin, M. Kihlgren, A. Rinell Hermansson and G. Östlinder (eds.), När omvårdnad blev vetenskap. De första decennierna [When caring became science. The first decades], Stockholm: Liber AB, pp. 328-337, 2014.

[14] H. Eriksson, "Vårdvetenskap för en postmodern tid [Caring Sciences for a Postmodern Era]," In Eriksson, H (ed.) Vårdvetenskap och postmodernitet [Caring Sciences and Postmodernity], Lund: Studentlitteratur, pp. 29-48, 2014.

[15] J. Y. Mattsson and L. Stevens, "Development of an individual assessment instrument for critical care nursing students," Journal of Nursing Education and Practice, vol. 7, no. 2, pp. 54-61, 2017.

[16] H. Engstrand and J. Mattsson, "The non-verbal communication in handover situations are the spice between the lines, to understand the severity of the patient´s condition," Journal of Nursing Education and Practice, vol. 7, no. 5, pp.1-8, 2017.

[17] G. Björling, D. Johansson, L. Bergström, S. Jalal, I. Kohn, C. Frostell and S. Kalman, S, "Tolerability and performance of BIP endotracheal tubes with noble metal alloy coating – a randomized clinical evaluation study," BMC Anesthesiology, vol. 15, no.174, 2015.

[18] G. Björling, D. Johansson, L. Bergström, J. Sanchez, A. Strekalovsky, C. Frostell and S. Kalman, "Central Venous Catheters Coated with a Noble Metal Alloy – A Randomized Clinical Pilot Study," Scand. Cardovascular Journal. In-press

[19] S. Frennert and B. Östlund, "What happens when seniors participate in new eHealth schemes?" Disability and Rehabilitation: Assistive Technology. July, pp-1-9, 2015.

[20] S. Frennert, H. Eftring and B. Östlund, "Case Report: Implications of Doing Research on Socially Assistive Robots in Real Homes," Int. JN of Social Robotics, Accepted January 11, 2017.

[21] B. Östlund, B. "Digital ageing in the making and its ethical challenges," In H. Campbell, G. Grieve and M. Lövheim (eds), Digital Existance, Routledge Studies in Digital Religion, In printing November 2016.

How to Evaluate an m-Health Project: Case of e-Ambulance in Japan

Masatsugu Tsuji
Faculty of Economics
Kobe International University
Kobe, Japan
e-mail: mtsuji@kobe-kiu.ac.jp

Yoshihisa Matsumoto
Graduate University for Advanced Studies
Tokyo, Japan
e-mail: ymatsumoto@avaya.com

Masaru Ogawa
Faculty of Business Management
Kobe Gakuin Universsity
Kobe, Japan
e-mail: CZK07133@nifty.ne.jp

***Abstract*— This study aims at evaluating the economic effect of a e-ambulance project, or emergency telemedicine in the rural areas in Kouchi Prefecture in Japan. The focus is on ambulances equipped with ICT devices which transmit images of acute patients to remote hospitals. Kouchi Prefecture started the e-ambulance project in Aki and Muroto Cities in 2012. From two cities, it takes approximately one hour to reach emergency hospitals located in Kouchi City, the prefectural capital. One of the merits of e-ambulance with the image transmitting system is that doctors in accepting hospitals can monitor real time situation of a patient and prepare for necessary treatment prior to the time patient arrives. They thus save time and effort. In measuring benefit, this study employs different methodology; the e-ambulance project enhances wellness of residents since they perceive more secure. Thus the contingent valuation method (CVM) is applied and willingness to pay (WTP) is used as an index of benefit and estimated based on surveys to residents, which amounts to 1,747 yen per resident per year. Total cost calculated is 381,792,228 yen over three years, and accordingly, B/C ratio amounts 0.459.**

Keywords- e-ambulance; WTP; Cost-benefit analysis; CVM; B/C ratio.

I. INTRODUCTION

e-Health has been implementing in many countries, and in order for the system to be diffused further, there are still lots of obstacles such as the legal framework, economic foundations of implementations, and other regulations. All medical systems were established in the age of face-to-face medicine fair prior to e-Health. To overcome these obstacles, it is required to demonstrate its effectiveness, that is, e-Health contributes to efficiency of medical services and enhances wellness of people [1]. Therefore, its cost-effectiveness must be proved by comparing its benefits and costs. The latter consists of equipment such as servers and peripheral devices, salaries and wages of doctors and nurses, and maintenance fees such as communication charges and other miscellaneous operating costs. On the other hand, to obtain its concrete benefits in monetary terms is analytically difficult, since the benefits mainly come from users' subjective satisfaction which is difficult to measure [2][3]. Without a firm basis of its cost-effectiveness, the future sustainability of the e-Health cannot be guaranteed.

The authors have been analyzing the benefits or merits of e-Heal in terms of the amounts of medical expenditures or treatment days saved by e-Health employing different methods, which are mainly statistical analyses [4]-[9]. These analyses became possible since the precise data of medical expenditures or treatment days were obtained. Based on the construction of the database, the effect of e-Health on these out comes were estimated for the various objectives or methods. The effect in the long-term effect such as 10 years is the example of the former [7], whereas the estimations without biases such as [5][8] are those of the latter. On the other hand, this study takes a different method other than statistical and focuses on the comparison of costs and benefits of e-Heath. In this sense, this paper examines e-Health in terms of cost effectiveness.

In order to measure the benefits of public services such as e-ambulance which are not traded on the market, the following methods are utilized in this field: (a) travel cost method: (b) replacement costs method; (c) hedonic approach; and (d) contingent valuation method (CVM). In what follows, this study employs CVM, which has been recently widely adopted in the fields of health economics and environmental economics [10][11][12]. CVM measures the benefits to users in terms of willingness to pay (WTP), which is the monetary amount that users want to pay for receiving the service. By asking WTP of each user, we can then construct the surrogate demand function for the e-Health service. It should be noted that this paper utilizes the results of economic evaluation of e-Health for the assessment of a project which aims to introduce the system. If the system is not introduced yet, it is impossible theoretically to evaluate a project by asking residents about WTP. CVM, however, can be applied even in this case, since it asks imaginary questions to people relating to benefits [13]. Although CVM and WTP have a strong theoretical basis, CVM tends to have a bias because it asks for concrete valuation and choice under fictitious circumstances. Care should be taken to clarify what kind of bias it possesses and to remove them.

This paper aims to analyze the Cost/Benefit analysis of the e-ambulance project in two cities in the rural and depopulated areas in Japan: Aki City and Muroto City, Kochi Prefecture. Ambulances are equipped with ICT devices which transmit images of patient to remote hospitals. In the depopulated areas, the number of clinics and medical specialist is small and a patient with acute disease or wounded by accident must be transported to hospitals with full facilities. Traditional ambulances are equipped with the mobile communication system only for voice or facsimile, and accordingly information transmitted from the ambulance to hospitals is limited. Kouchi Prefecture started the e-ambulance project in Aki and Muroto Cities in 2012. From these two cities, it takes about approximately one hour to reach emergency hospitals located in Kouchi City, the prefectural capital. One of the merits of e-ambulance with the image transmitting system is that doctors in the accepting hospital can monitor real time situation of a patient and prepare for necessary treatment when patient arrives. They thus save time and effort.

The analysis in this study provides not only the firm theoretical basis of evaluating e-Health project but also practical guideline for regions which plan to implement e-Health. In particular, surveys to residents were conducted not only above two cities but also Ino Town, Kochi Prefecture, which plans to implement e-ambulance. In this context, we can compare how WTP is different between two types regions which are implementing or not, we can compare ex-post and ex-ante WTP. This study therefore leads to useful information to local governments which plan to implement e-Health.

The structure of the paper is as follows: In the next section, the surveys to residents and their results are briefly summarized and explained. Section III estimates WTP based on the data. The content of various costs and the comparison with benefits are analyzed in Section IV. The brief conclusion and further analysis are presented the final section.

II. SURVEYS CONDUCTED

A. *Aki and Kuroto Cities*

In this section, let us briefly describe Aki and Muroto cities. These neighboring cities are located in east Kochi Prefecture. They are mountainous and face the Pacific Ocean. Their primary industries are agriculture and fishery. Their population is declining, while the percentage of the elderly is increasing. Aki City with the area of 317.37km^2 has 18,657 residents and 8,055 households, and the elderly ratio is 29.14%, while Muroto with the area of 248.25 km^2 has population 17,490 and 7,598 households, and its elderly ratio is 32.94%. There are 42 in Aki City and 23 clinics in Muroto City, but no tertiary emergency medical facility in two cities and therefore patients who emergency services need have to be transferred to Kochi City. This is the background of the e-ambulance project.

B. *Surveys to residents*

The surveys were conducted to residents in Ino Town on November 5, Aki City on November 18 and Muroto City on November 19, 2013. We interviewed 62 in Ino Town, 55 Aki City, and 47 Muroto City, totaling 164, and asked questions pertaining to the following: (a) WTP; (b) effectiveness; (c) frequency of usage; and (d) user properties such as age, gender, income, education, and health condition. These are supposed to affect WTP of residents.

Let us examine characteristics of residents who replied to our questionnaires. 76 are males, while 85 are females (TABLE I). The average age is 42.7 (TABLE II) and regarding education, high school, junior college, and university and higher are equally distributed (TABLE III). 153 are still working, while 8 not working (TABLE IV). The average family size is 2.9 (TABLE V), and people living alone and living with wife or husband share more than half. The number of children and grandchildren living in the same city is shown in TABLE VI. Regarding health condition, more than two-thirds replied either good, fair or all right (TABLE VII). Accordingly, the average frequency of visiting medical institutions per month is 1.6, and more than 90% reported no necessity for outside medical help, and these are due to their average ages (TABLE VIII). Even though two cities located depopulated areas, they are enough neighboring clinics, and the most of them live close to clinics (TABLE IX).

TABLE I. GENDER

	Freq.	%
1 Male	76	47.2
2 Female	85	52.8
Total	161	100

TABLE II. AGE DISTRIBUTION

Age	Freq.	%
20-24	12	7.5
25-29	13	8.1
30-34	6	3.8
35-39	16	10.0
40-44	17	10.6
45-49	24	15.0
50-54	29	18.1
55-59	22	13.8
60-64	10	6.3
65-69	6	3.8
70-74	4	2.5
75-79	1	0.6
total	160	100.0

TABLE III. EDUCATION

	Freq.	%
1. Junior high school	5	3.1
2 High school	56	34.8
3 Junior collage	51	31.7
4 University and higher	49	30.4
total	161	100.0

TABLE IV. EMPLOYMENT

	Freq.	%
1 Working	153	95.0
2 Not working	8	5.0
total	161	100.0

TABLE V. NUMBER OF FAMILY

	Freq.	%
0	15	11.1
2	27	20.0
3	38	28.1
4	0	0.0
5	12	8.9
6	3	2.2
total	135	100.0

TABLE VI. NUMBER OF CHILDREN AND GRANDCHILDREN LIVING IN THE SAME CITY

	Freq.	%
0	123	76.4
1	14	8.7
2	9	5.6
3	9	5.6
4	3	1.9
5	2	1.2
6	1	0.6
total	161	100.0

TABLE VII. DISEASES TREATED

Diseases	Freq.	%
1 High blood pressure, Atherosclerosis	11	27.5
2 Heart diseases	1	2.5
3 Diabetes	2	5.0
4 Stroke	1	2.5
5 Chronic Gastritis, Gastric ulcer	3	7.5
6 Assume	1	2.5
7 Backache, arthritis, rheumatism	6	15.0
8 Gglaucoma, cataract	1	2.5
9 Rrenal disease, kidney failure	3	7.5
10 Hemorrhoid	0	0.0
11 Others	11	27.5
total	40	100.0

TABLE VIII. FREQUENCY OF VISITING CLINIC

Number	Freq.	%
1 0	90	57.3
2 1-2	66	42.0
3 3-5	1	0.6
4 6-9	0	0.0
5 more than10	0	0.0
total	157	100.0

TABLE IX. MINUTES TO THE NEAREST CLINIC

Time	Freq.	%
1 less than 10 minutes	79	50.6
2 10-less than 30	52	33.3
3 30-less than 60	16	10.3
4 60-less than 2 hours	9	5.8
total	156	100.0

III. Estimation of WTP

A. CVM

In order to measure the benefits of services which are not traded on the market, the following methods are utilized: (a) travel cost method: (b) replacement costs method; (c) hedonic approach; and (d) CVM. In what follows, we use CVM, which has been recently widely adopted in the fields of health economics and environmental economics [10][11][12]. CVM measures the benefits to residents or users of e-Health in terms of WTP, which is the monetary amount which users are willing to pay for receiving the service. By asking the WTP of each user, we can then construct the surrogate demand function for the e-ambulance system. Although CVM and WTP have a strong theoretical basis, CVM tends to have a bias because it asks for concrete valuation and choice under fictitious circumstances. Care should be taken to clarify what kind of bias it possesses and to remove them.

B. Questionnaire

We conducted the surveys to residents in Ino Twon on 5, Aki City on 18 and Muroto City on 19, November 5, 2013, and asked questions pertaining to the following: (a) WTP; (b) whether they know the e-ambulance project, (c) desire to continue the project; and (d) user properties such as age, gender, income, education, and health condition. The questionnaire related to WTP is based on the three-stage double bound method: We begin by asking whether they would be willing to pay monthly charges of 1,500 yen (US$15). This initial value in CVM method is important, since WTP tends to depend on the initial value. If their answer is "yes," we then ask whether they would be willing to pay 2,500 yen (US$25). If they reply "yes" again to 2,500 yen, their WTP is 2,500 yen. If "no", then we lower the amount to 2,000 yen (US$20). If they reply "yes" to 2,000 yen, then that is their WTP. If again their answer is "no," we lower the amount further to 1,500 yen. In the first question of 1,500, if the reply is "no" to 1,500 yen, then we lower the amount to 500 yen. If the reply to 500 yen is "yes," then we ask whether 1,000 yen is acceptable. If the reply to 1,000 yen is "yes," then his/her WTP is 1,000 yen. If not, it becomes 500 yen. On the other hand, the reply to 500 yen is "no," then we ask how about 250 yen. If the reply is "yes," then WTP becomes 250 yen. If it is "no," then we ask how much he/she wants to pay. They reply their acceptable amounts. These series of questions are standard in the evaluation of issues in public services, environments, and so on.

The distribution of WTP from the survey shown in TABLE X is as follows: more than 3,000 yen (5), 2,500-2,999 yen (5), 2,000-2,499 yen (5), 1,500-1,999 yen (10), 1,000-1,499 yen (18), 500-999 yen (30), 250-499 yen (36), and 1-249 yen (11). It should be noted that there are 33 residents responded that their WTPs is zero. After checking their reasons, those are considered as "non-response," and their replies are omitted from the analysis.

TABLE X. DISTRIBUTION OF REPLIED WTP

WTP (JPY)	3,000>	2,500-2,999	2,000-2,499	1,500-1,999
Number	5	5	5	10
WTP (JPY)	1,000-1,499	500-999	250～499	1～249
Number	18	30	36	11

C. *Estimation of demand function and WTP*

Based on the above WTP of each respondent, the demand function of the e-ambulance service is estimated; more precisely, the probability of acceptance to amounts questioned is estimated and the number of residents who will agree to pay. The functional form of demand to be estimated is assumed to be logistic, namely,

$$\text{Probability of acceptance} = 1 - 1/(1 + \exp(-\alpha - \beta \log WTP)) \quad (1)$$

The probability of acceptance is the ratio of the residents who reply that they are willing to use the device at the amount of charges provided in the questions. The estimated coefficients α and β are summarized in TABLE XI. The average WTP is calculated as the area under this demand function, which results in being 1,747 yen (approximately US$175) per resident per year. The mean value For WTP, which is the amount that the probability of acceptance is 50%, is estimated at 1,379.45 yen. This paper uses the average value as WTP in the analysis in what follows.

TABLE XI. RESULTS OF ESTIMATION

	Coefficients	S.D.	t-value	p-value	
α	18.765	1.276	14.711	0.000	***
β	2.596	0.175	14.854	0.000	***

Log likelihood function: -475.7578

IV. COST BENEFIT ANALYSIS

A. *Total benefits*

In the Cost/Benefit Analysis, total benefits and costs are compared over the period of several years. In this paper, two kinds of the time span are used: one is three years, which is the same as the project period, while that of five year is also considered, which is usual period of public projects like e-ambulance. WTP obtained above is for per resident per year, and it is multiplied by total number of residents, since all residents have a chance to use ambulance. The population of each city is 18,657 in Aki City and 17,490 in Muroto City as of January 1, 2014, and thus total population is 36,147. Since all residents have possibility of using ambulance, even if they are babies or 100 years old, the numbers of residents who receive benefit are total of two cities, we multiply WTP 1,747 yen by population of 36,147, which ends up with 63,148,089 (US$631,000). That is, one-year benefits of e-ambulance services total approximately 63,148,809 yen (US$631,000). In order to obtain three and five years' worth of benefits, the present values of three or five years' benefits are calculated with a 4% discount rate, and we assume that population of two cities remains at the level for six years. This results in three (five) years' benefits totaling 175,243,694 yen (281,127,278 yen)

B. *Total costs*

The total cost of the system consists of two major categories; initial fixed and annual operating costs. The former is the items which have to pay at the first year of the project and covers that (i) ICT hardware equipment of the systems of transmitting and receiving images and related equipment, (ii) ambulance, (iii) costs related to software development and the purchase of software, (iv) installment, and (v) initial training cost. The latter, on the other hand, the latter is required annually and contains the followings: (vi) salary of ambulance crew; (vii) maintenance fees which consist of those related to hardware and software; (viii) gasoline mainly for ambulance; and (ix) communications charges for the wireless and wired devices. These cost items are summarized in TABLE XII for the annual basis.

TABLE XII. COSTS OF THE E-AMBULANCE PROJECT

C. Total Cost (single year)	JPY
1. Initial cost	
Hardware (equipment)	
1-A Equipment	8,144,662
Ambulance	
1-B Ambulance	36,000,000
Software	
1-C Software development	19,341,000
1-D Software	4,114,803
Installment	
1-E Installment cost	2,706,784
Training	
1-F Initial training	1,680,000
subtotal	71,987,249
2. Operational Cost	
Salary	
2-A Salary of ambulance crew	50,572,080
Maintenance	
2-B Software maintenance	893,928
2-C Hardware maintenance	654,360
Fuel	
2-D Gasoline	2,400,000
Communication	
2-E Communication fees	1,200,000
subtotal	55,720,368
Total	127,707,617

In order to obtain total costs of three (five) year period, operational annual costs must be discounted at a 4% discount rate. As a result, the costs of salary in three (five) years are 140,342,126 yen (225,137,915 yen), respectively, while those of miscellaneous expense including maintenance, gasoline and communications are 9,990,328 yen (16,026,560 yen). Thus, total operating costs for three (five) years are 150,332,453 yen (241,164,475 yen), respectively.

C. B/C ratio

From the above calculation, benefits total 175,243,694 yen (US$1,752,436) over the period of three years and 281,127,278 yen (US$2,811,272) over the period of five years, whereas costs total 381,792,228 yen (US$3,817,922) over the period of three years and 472,624,250 yen (US$4,726,242) over the period of five years. On the other hand, total costs amount to 381,792,228 yen (US$3,817,922) for three years and 472,624,250 (US$4,726,242) yen for five years. Therefore, the B/C ratio over the period of three years is 0.459 (see TABLE XIII), while 0.595 over the period of five years, that is, benefits are about half of costs for three year project, and about 60% for five year project (see TABLE XIV). It can be concluded that benefits are far smaller than costs.

Here the viewpoint is changed, that from the view of local governments which implementing the project, that is, Aki and Muroto Cities, they only bear the costs of operating costs, since initial costs are borne by subsidies from the central government, and they can bear only cost of operating costs. The B/C ratio calculated in this way is 1.166 over the both of periods indicating that for two local governments, benefits exceed its costs. Thus from the view of city, this project is favorable, and worthy to implement.

TABLE XIII. COST/BENEFIT: 3 YEARS

	JPY
Total benefit B3	175,243,694
Initial cist IC3	231,459,775
Operating cost OC3	150,332,453
Total Cost TC3	381,792,228
B3/TC3	0.459
B3/OC3	1.166

TABLE XIV. COST/BENEFIT: 5 YEARS

	JPY
Total benefit B5	281,127,278
Initial cist IC5	231,459,775
Operating cost OC5	241,164,475
Total Cost TC5	472,624,250
B5/TC5	0.595
B5/OC5	1.166

D. Comparison with other e-Health projects

There is no economic evaluation of e-ambulance, but the results can be compared with other e-Health projects which are implementation of telecare. Telecare transmits health-related data of its users such as blood pressure, ECG, and blood oxygen to a remote medical institution via a telecommunications network. The system is equipped with a simple device which, when used continuously, records the condition of the elderly or a patient's illness in graphs, which are then used for diagnosis and consultation. Reports sent by the medical institution are also helpful for users to enhance their daily health consciousness and make an effort to maintain good health. Such positive effects have been identified through field surveys.

Tsuji et al. [14] applied the Cost/Benefit Analysis to the telehealth systems in the following four regions in Japan: Kamaishi City, Iwate Prefecture; Nishiaizu Town and Katsurao Village, Fukushima Prefecture; and Sangawa Town, Kagawa Prefecture. Benefits are expressed in terms of WTP based on CVM, whereas the costs are calculated as the sum of equipment, salaries of doctors and nurses, and other operations. Then, the benefits and costs are compared in terms of the B/C ratio, and the results obtained are shown in TABLE XV. Since the users receive and perceive benefit from telecare by using every day, while benefits of e-ambulance are less noticed by residents, WTP of e-ambulance tends to be larger than e-ambulance.

TABLE XV. COSTS AND BENEFITS OF TELECARE BY CVM

Unit: JPY

	Kamaishi	Nishiaizu
Number of devices	211	400
WTP	4,519	3,177
Equipment	39.9*	136.7*
Salaries	8.6*	3.7*
Others	1.9*	1.9*
Total costs (6 years)	95.5*	184.5*
B/C	1.07	0.58
(B/C)**	1.87	2.31
	Katsurao	Sangawa
Number of devices	325	225
WTP	1,640	2,955
Equipment	111.4*	133.5*
Salaries	3.36*	4.5*
Others	10.4*	3.0*
Total costs (6 years)	184.2*	174.3*
B/C	0.54	0.61
(B/C)**	1.42	2.60

Note 1: * indicates million yen

Note 2: (B/C)** indicates Benefit/Operating cost

Source: [14]

Thus the B/C ratios obtained for e-ambulance are similar to those of telecare, which were operated by local governments and received subsides for initial equipment from the central government. As a project, they are less than 1, that is, benefits are smaller than costs, while for local governments it is worthy to implement since benefits to the users are larger than costs which were borne by local governments. From the view of local governments which implementing the project, that is, Aki and Muroto Cities, they only bear the costs of operating costs, since initial costs are borne by subsidies from the central government, and they can bear only cost of operating costs. The B/C ratio calculated in this way is 1.166

over the both of periods indicating that for two local governments, benefits exceed its costs. Thus from the view of city, this project is favorable, and worthy to implement.

V. CONCLUSIONS

In this paper, WTP in Aki and Kuroto Cities and Ino Town is estimated by CVM and WTP obtained is 1,747 yen. According to our rigorous analysis, we found that this value is not different from the ex-post WTPs estimated in our previous research. The effects of the e-ambulance in Aki and Muroto Cities are also similar to realized ones in the other regions. These results indicate that WTP can be an indicator of potential effectiveness of regional health policy implemented by local governments.

So far, we have conducted surveys of four local governments as shown in TABLE XV. Except for Kamaishi City, their B/C ratios are approximately 0.5, that is, benefits cover only half of the costs. In addition, regarding the frequency of usage of the device, Kamaishi City also has a much higher ratio than the other local governments. This is due to charges, not free like other region, their efforts to promote usage such as a users' association which organizes events to enhance consciousness towards health, and the participation by medical doctors in this system, which increases the users' reliance on the system. It is clear from our previous studies that telecare is useful for consultation and maintaining the good health of the elderly and patients suffering from chronic diseases who are in stable condition, but it is not for curing disease. It therefore has a psychological effect such as providing a sense of relief to its users by the knowledge of being monitored by a medical institution 24 hours a day. This makes it difficult to estimate its benefits in concrete terms.

On the other hand, benefits of e-ambulance are hardly perceived by residents, except transported by ambulance. But the residents feel wellness because of e-ambulance, since in case of acute diseases they would be treated better than the situation without it. These benefits are less perceived and it is difficult to measure. CVM seems to be the only suitable method to measure benefits.

REFERENCES

[1] M. Tsuji, "The Telehomecare/telehealth System in Japan," Business Briefing: Global HealthCare, pp. 72-6, London, 2002.

[2] M. Tsuji, W. Suzuki, and F. Taoka, "An Empirical Analysis of a Telehealth System in term of Cost-sharing," Journal of Telemedicine and Telecare, Vol. 9, pp. 41-43, 2003

[3] M. Tsuji and W. Suzuki, "The application of CVM for assessing the tele-health system: an analysis of the discrepancy between WTP and WTA based on survey data," (with Wataru Suzuki) in Assets, Beliefs, and Equilibria in Economic Dynamics: Essays in Honor of Mordecai Kurz, ed. by C. D. Aliprantis, K. Arrow, P. Hammond, F. Kubler, H.-M. Wu, N. Yannelis, pp. 493-506, Springer, 2003.

[4] Y. Akematsu and M. Tsuji, "Estimating the Effect of eHealth on Medical Expenditure," Journal of Telemedicine and Telecare, Vol. 16, pp. 169-71, 2012.

[5] Y. Akematsu and M. Tsuji, "Measuring the Effect of Telecare on Medical Expenditures without Bias Using the Propensity Score Matching Method," Telemedicine and e-Health, Vol. 18, pp. 743-747, 2012.

[6] Y. Akematsu, and M. Tsuji, "Relation between telecare implementation and number of treatment days in a Japanese town," Journal of Telemedicine and Telecare, Vol. 19, pp. 36-39, 2013.

[7] Y. Akematsu, S. Nitta, K. Morita, and M. Tsuji, "Empirical analysis of the long-term effects of telecare use in Nishi-aizu Town, Fukushima Prefecture, Japan," Technology and HealthCare, vol. 21, pp. 173-82, 2013.

[8] K. Minetaki, Y. Akematsu, and M. Tsuji, "Effect of e-Health on Medical Expenditures of Outpatients with Lifestyle-related Diseases," Telemedicine and e-Health, Vol. 17, pp, pp. 591-5, 2011.

[9] K. Minetaki, Y. Akematsu, and M. Tsuji, "Empirical Study of Emergency Medical Service," Journal of eHealth Technology and Application, Vol. 18, pp. 167-76, 2010.

[10] G. Garrod, and K. G. Willis, Economic Valuation of the Environment: Method and Case Studies, Edward Elgar, London, UK, 1999.

[11] J. B. Loomis, and R. G. Walsh, Recreation Economic Decisions: Comparing Benefit and Costs, Second Edition, Venture Publishing Inc., State College, Pennsylvania, 1997.

[12] M. Tsuji, W. Suzuki, "An Economic Assessment of Tele-health: The WTP Approach," Proceedings of APT Conference, Jakarta, Indonesia, 2002.

[13] M. Tsuji, C. Iizuka and, F. Taaoka, "On the Ex-ante and Ex-post Evaluation of Economic Benefits of the Japanese e-Health System," Proceedings of 3rd APT Telemedicine Workshop 2005, pp. 68-72, Multimedia University, Kuala Lumpur, Malaysia, 2005.

[14] M. Tsuji, W. Suzuki, and F. Taoka, "An Empirical Analysis of the Economic Assessment of the Tele-health System by CVM," in Proceedings of MODSIM2003, pp. 1950-5, Townsvill, Australia, 2003.

Proposal of Gait Monitoring Shoes Based on Comparative Analysis of Walking Gait Cycle between Normal People and Stroke Patients

Shohei Yoshida
Graduate School of Software and Information Science
Iwate Prefectural University Graduate School
Takizawa, Japan
e-mail: g231n034@s.iwate-pu.ac.jp

Yoshitoshi Murata
Faculty of Software and Information Science
Iwate Prefectural University
Takizawa, Japan
e-mail: y-murata@iwate-pu.ac.jp,

***Abstract*— Elderly people have an increased risk of falling and consequently injuring themselves. The prevention of falls is one of major important issues to keep healthy life, because a fall is a main reason for them to hospital administration or a residential care. Motion capture systems are a key component to prevent falls. We comparably analyzed walking gait cycle between normal people and stroke patients using a wearable device (WD) to detect warning signs of falls. In this paper, we experimentally clarify to which body part a WD is best attached and what kind of signs are useful to prevent falls. We also develop a pair of shoes with a WD in each heel in accordance with results of pre-experiment.**

Keywords-falls; trip; wearable device; hemiplegia; guit; shoes.

I. INTRODUCTION

Percentage of elderly people in the population is increasing in the world [1]. As the number of elderly increases, the number of functionally impaired people, such as cerebrovascular patients who have one side paralysis, will also increase. These people have an increased risk of falling and consequently injuring themselves [2]. Falling down is one if the main reasons for them to be hospitalized or placed in residential care.

There are many researches related to a fall for elder people. The World Health Organization Regional Office for Europe analyzed these studies and classified fall risks amongst elderly people by history of falls, age, gender, living alone, ethnicity, medicines, medical conditions, impaired mobility and gait, sedentary behavior, psychological status - fear of falling, nutritional deficiencies, visual impairments, and foot problems [2]. Stroke patients, such as a cerebrovascular disease especially are at substantially high risk of falling [3][4][5][6]. The higher frequency of falls for them due to weak muscles, one-side paralysis, and toes pointing down. For people with impaired mobility and gait, tripping is a major cause of falls [7][8], so we focus on tripping in this paper.

Since weak muscles, one-side paralysis, and toes pointing down strongly appear in the movement of leg and foot, motion capture for them are a key component to analyze impaired mobility and gait and useful to prevent tripping and conduct therapy and rehabilitation of hemiplegia.

Weijun Tao et al. reviewed gait analysis technologies based on wearable sensors that were the accelerometer, gyroscope, electromagnetic tracking system, magneto-resistive sensors, flexible goniometer, sensing fabric, force sensor, and sensors for electromyography. They mentioned that fall risk estimation is an important application of gait analysis using wearable sensors [9]. However, they did not describe about motion of gait for elderly people or stroke patients.

Stacy J. Morris Bamberg et al. developed a prototype shoe in which several kinds of wearable sensors, such as accelerometer, gyroscope, force sensor, bidirectional bend sensor and so on [10]. The calibrated sensor outputs were almost same as results obtained simultaneously from a biological motion measuring equipment. They calculated the maximum pitch (angle between the shoe sole and floor at the toe-off timing), minimum pitch (angle between the shoe sole and floor at the heel-strike timing), the stride length from output of accelerometers and gyroscopes. They also compared the maximum pitch, minimum pitch and stride length between the healthy gait and parkinsonian gait. There were differences on mean value of calculated data between the healthy gait and parkinsonian gait. However, considering standard deviation of calculated data, such differences were small. They also did not measure and analyze motions of gait for elderly people or stroke patients.

Farzin Dadashi et al. measured motion of gait for many elderly people with shoe-worn inertial sensors and provided normative values for a clinician to measure reference gait parameters [11]. They analyzed motion of gait and clarified the difference in gait parameters, such as the clearance between a shoe sole and floor, gait speed, stride length between males and females by considering the effect of age factors. However, their data did not show differences clearly between the male and female, and the effect of age factor. And, they did not investigate data for stroke patients or analyze reasons for tripping.

Mourad Benoussaad et al. introduced a method to robustly estimate foot clearance during walking using a single inertial measurement unit (IMU) placed on the subject's foot [12]. In their paper, the foot clearance was the height of ankle from a floor. However, the toe clearance is more critical for tripping. And, they did not measure the toe clearance for stroke patient or analyze reasons for tripping.

Here, we focus on extracting warning signs of tripping for stroke patients, such as cerebrovascular patients. We got output data of an acceleration sensor and gyroscope sensor in the wearable device (WD), Sony SmartWatch 3, mounted on the front part of a foot to estimate the kicking power and change of angle between a foot and floor. We noticed that an angle velocity at the terminal stance and an angle at the terminal swing show clearly differences between unimpaired student and stroke patients. After that, we develop a pair of shoes built a WD on.

In the next section, we consider how people trip on a flat floor. Different features between physically unimpaired students and stroke patients are extracted from measured data in Section III. The best body part to which a wearable device (WD) can be attached is described in Section IV. We introduce a pair of shoes with a WD in each heel and example data measured using them in Section V. Conclusions are summarized in Section VI.

II. CONSIDERATION OF TRIPPING FACTOR

When the swing foot progression is unexpectedly obstructed, a trip occurs that leads to a forward rotation of the body and eventually might cause a fall.

Mourad Benoussaad et al. measured the minimum toe clearance (MinTC) to avoid tripping. MinTC is a critical value to clear obstacles on the ground or floor. However, elderly people, especially those who have had strokes, sometimes trip on flat ground or floors, not obstacles. In this section, we consider reasons a person trips on flat ground or floors. We divide the normal walking gait cycle into eight phases the same as Weijun Tao et al. as shown in Fig. 1 [9]: (1) initial contact (heel-strike timing), (2) loading response, (3) mid-stance, (4) terminal stance (toe-off timing), (5) pre-swing, (6) initial swing, (7) mid-swing, and (8) terminal swing.

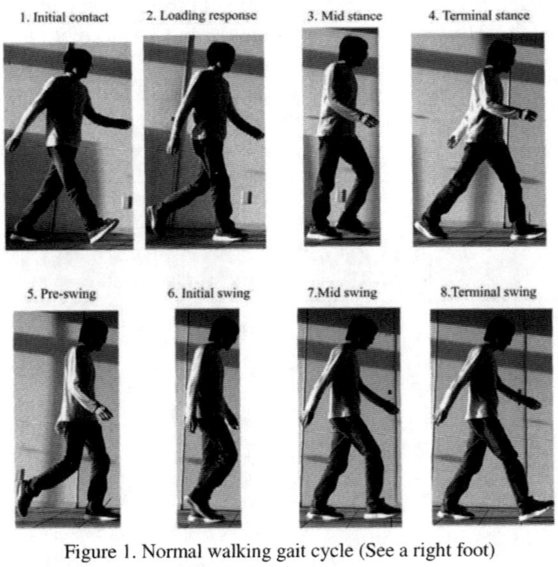

Figure 1. Normal walking gait cycle (See a right foot)

Most stroke patients have weak muscles and are hard to raise their toe. Therefore, they are at risk of three types of trips.

- Case 1: A toe touches the floor first instead of a heel at phase 1. Since phases 2-5 are skipped, the toe is dragged along the floor. When the dragging strength is stronger than the person's muscular power, he/she trips (Fig. 2(a)).
- Case 2: Kicking power of the front part of a foot is insufficient at phases 4 and 5 to raise the heel and toe up from the floor. In this case, a person does not swing but shuffles. When the frictional force between a shoe sole and the ground or floor is stronger than his/her muscular power, he/she trips (Fig. 2(b)).
- Case 3: A toe touches the floor due to it pointing down during the swing phases (5-8), and the knee goes further forward than the foot. When the dragging strength is stronger than the person's muscular power, he/she trips (Fig. 2(c)).

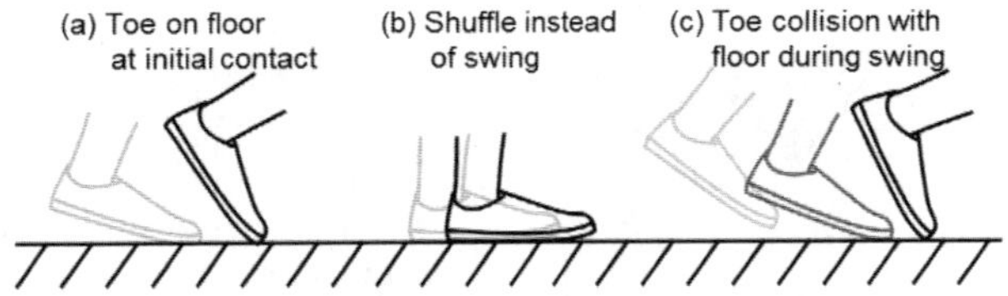

Figure 2. Cases of tripping

The above problems suggest that the kicking power at phases 4 and 5 and the angle between the foot and a floor are critical parameters.

III. EXTRACTION OF WARNING SIGNS FOR TRIPPING

In this section, we experimentally investigate whether the kicking power at phases 4 and 5, and the angle between the foot and lower limb are critical parameters.

A. Experimental method

Since kicking power must be expressed as the angle velocity or the acceleration for the foot, we mounted a WD which had an accelerometer and gyroscope on the foot. In this experiment, we used Sony SmartWatch 3 as a WD which is mounted on the front part of a foot with Velcro tape as shown in Fig. 3. This mounting position was same as one in Farzin Dadashi's experiment [11]. The sampling rate was 40 msec.

We measure angle velocity for up and down directions of the front part of the foot (X axis of a 3D gyroscope). We also adopted a three-point moving average of the angle velocity to calculate the angle, because output values extremely change up and down. Therefore, the angle for X axis $Angle_{xn}$ at time t_n is calculated as follows.

$$Angle_{xn} = Angle_{xn-1} + \frac{t_n - t_{n-1}}{1000} \times \frac{G_{n-1} + G_n + G_{n+1}}{3} \quad (1)$$

G_{xn} is the value of angle velocity for X axis at time t_n. We also recorded video and measured locations of a patient's knee, ankle and foot by using Microsoft-KINECT to monitor their motions (Fig. 4). The UNIX time is introduced to synchronously measure data with a WD and MS-KINECT.

Before measuring, we investigated the measuring accuracy of Sony SmartWatch 3 using a slant rule as shown in Fig. 5. We measured data five times. Calculated angles vs. angles given by the slant rule are listed in Table I. These data showed calculated angles were so accurate. We noticed drift errors of a gyroscope that increase the value by 0.2 rad. /sec. during a WD sets on a flat floor. However, each measurement lasted less than 20 sec. Therefore, we think the effect of the drift error is negligible.

Participants were five physically unimpaired students and five stroke patients. Every stroke patients in this experiment had one-side paralysis, and trained periodically at a rehabilitation facility. Some of them used a wheel chair and could not walk by himself before training. They walked along a straight line to MS-KINECT. A WD was attached on the front part of foot on the paralysis side. We measured data for each stroke patient two times.

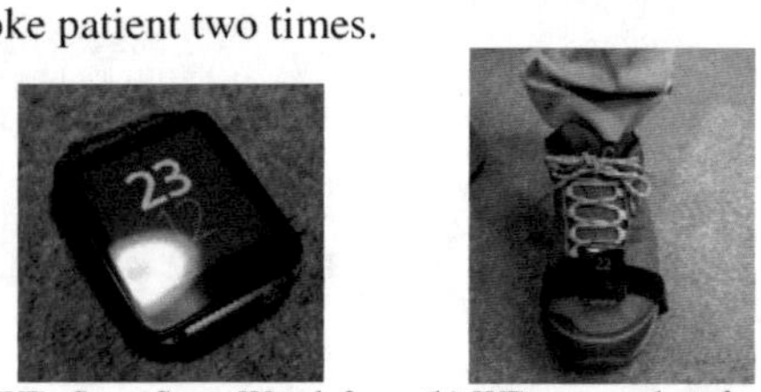

(a) WD: Sony SmartWatch 3 (b) WD mounted on foot
Figure 3. Measuring device and WD mounting method

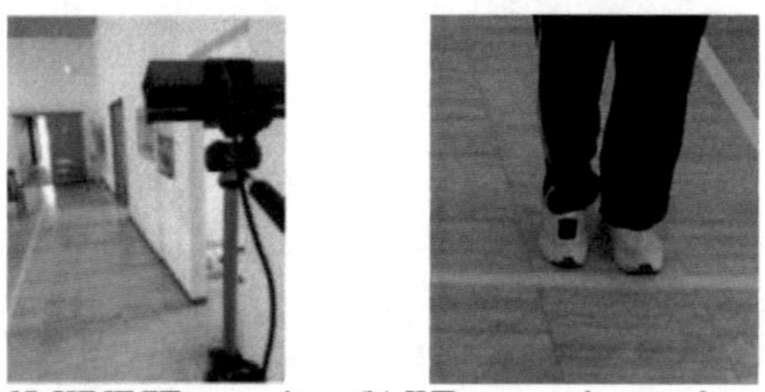

(a) MS-KINECT set up in rehabilitation facility (b) WD mounted on stroke patient
Figure 4. Measurement environment in rehabilitation facility

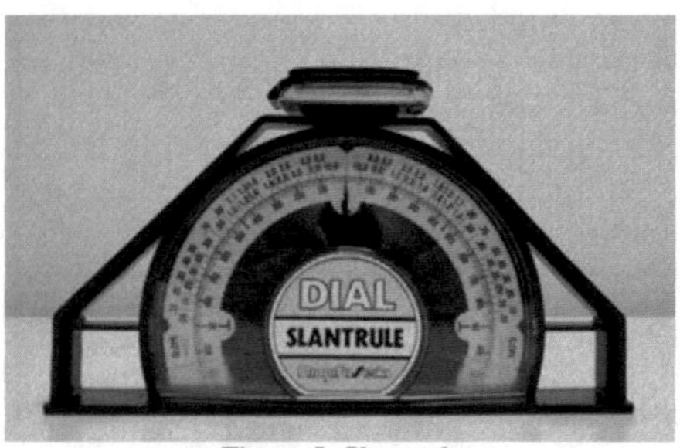

Figure 5. Slant rule

TABLE I. ACCURACY OF CALCULATED ANGLES

Given angle (degree)	Calculated angle (degree)	SD (degree)
+50	+49.00	0.45
+40	+39.17	0.74
+30	+28.92	0.51
+20	+19.58	0.63
+10	+9.21	0.58
0	0.21	0.15
-10	-9.72	0.58
-20	-19.47	0.34
-30	-30.70	0.56
-40	-40.57	0.41
-50	-50.54	0.53

B. Measured data and consideration

Fig. 6 and 7 show examples of change of angle velocity, angle, and acceleration for a physically unimpaired student and a stroke patient. Data for two steps are plotted.

Each flat period (roughly the center period) in these figures is when the entire shoe sole touched the floor; this period corresponds to phases 2 (loading response) and 3 (mid-stance). The reason that the value during this period is not zero is that the WD measures the angle between the front part of the foot and the floor, which depends on the person and shoe. Therefore, we reset this angle for the gait monitoring shoes described in Sections IV and V when the entire shoe sole touched the floor. This processing enables the WD to measure the angle between the back of the foot and the floor. This value does not depend on person or shoe.

The maximum angle velocity at timing A means the kicking power from phase 4 (terminal stance) to 5 (pre-swing), and the minimum angle at timing B means the angle to the floor at phase 8 (terminal swing). This processing removes the drift error of the gyroscope.

Lower angle velocity at A in Fig. 6 is about 420 deg./sec. On the other hand, higher angle velocity at A in Fig. 7 is about 250 deg./sec. Thus, a physically unimpaired student and a stroke patient obviously differ in terms of gait. The stroke patient clearly has weaker kicking power at phase 4 (terminal stance) than the physically unimpaired student.

Higher angle at B in Fig. 6 is about -18 degree. On the other hand, lower angle at B in Fig. 7 is about -8 degree. Thus, a physically unimpaired student and a stroke patient obviously differ in terms of the angle to a floor at phase 8 (terminal swing). This shows that it is difficult for a stroke patient to raise his or her toe at the terminal swing phase.

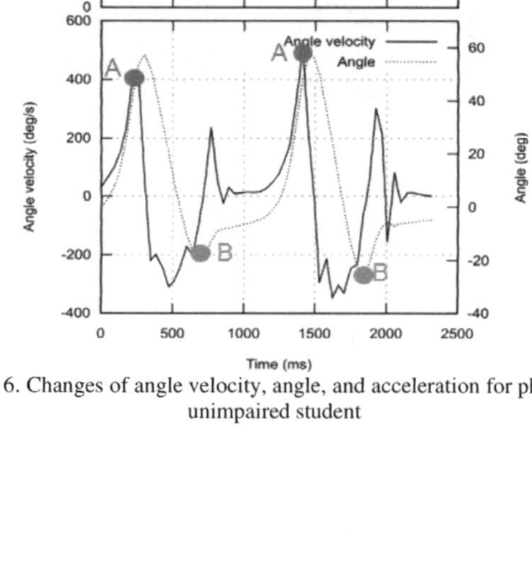

Figure 6. Changes of angle velocity, angle, and acceleration for physically unimpaired student

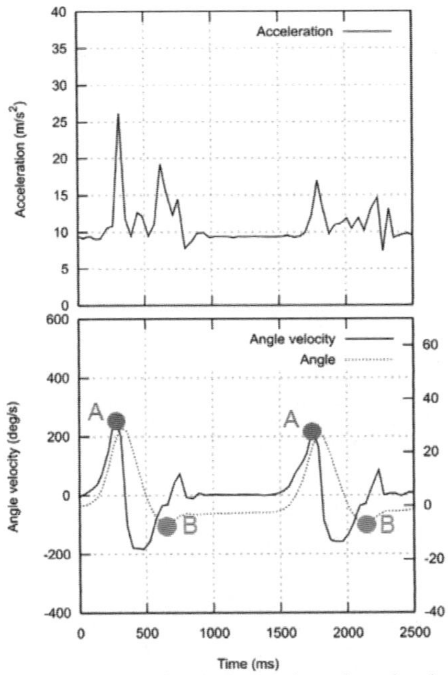

Figure 7. Changes of angle velocity, angle, and acceleration for stroke patient

The other hand, the acceleration basically change corresponding to the angle velocity and angle. However, they have much noise, and their amplitudes are not stable.

Tables II and III list the averages and standard deviations (SDs) of measured data for angle velocity at timing A and angle at timing B. The angle velocity at timing A is clearly different between unimpaired students and stroke patients. There is a big difference between them in the angle at timing B, however this value would be sometimes overlapped each other.

We also measure the cadence for a gait using the WD and MS-KINECT. In this paper, we define the cadence as the number of steps per minute. We estimated the cadence derived from an average of 10 intervals between one timing A and the next A, which are peak angle velocities of a step, when a WD is used. Estimated cadence is listed in Table IV. A physically unimpaired student and stroke patient do not obviously differ in terms of cadence.

As the result of this experiment, we decided to adopt the change of angle velocity and angle between a foot and floor to detect warning signs of falls.

TABLE II. ANGLE VELOCITY AT THE TERMINAL STANCE

Participant	Average (deg./s)	SD (deg./s)
Student	509.36	18.91
Stroke patient	342.06	86.52

Table III. Angle at the terminal swing

Participant	Average (deg.)	SD (deg.)
Student	-17.76	8.02
Stroke patient	-7.45	8.02

TABLE IV. ESTIMATED CADENCE

Participant	Average (steps/m)	SD (steps/m)
Student	46.4	5.0
Stroke patient	49.0	6.2

IV. SUITABLE WD ATTACHING POSITION

A WD has to be attached somewhere on a body during walking to detect signs of tripping to prevent a fall. A WD was attached to the front part of the foot in Section III. However, it is difficult for a WD to firmly be set at this place for a long time because it is easily detached. Therefore, we studied which position is the best to detect the change of angle velocity for a foot and angle between a foot and floor. We attached WDs to a heel and a lower limb as shown in Fig. 8.

For this test, we used STEVAL-WESU1 by STMicro-electronics (See Fig. 9) as a WD instead of Sony SmartWatch 3. This wearable unit includes four sensors:

- 3D-accelerometer,
- 3D-gyroscope,
- 3D-magnetometer,
- MEMS pressure.

This device is 37 x 40 x 8 mm and weighs 9.6 g.

We inserted STEVAL-WESU1 into the heel of a shoe as shown in Fig. 8 (a) (details in Section V). Angle velocity and acceleration data of STEVAL-WESU1 are sent to and processed by an Android smartphone. The sampling rate was 40 ms. We adopt a three-point moving average to remove noise.

We requested three unimpaired students to walk with their normal gait. Since their data change was basically the same, graphs of one participant are shown in Fig. 10 and 11. Both plotted lines in Fig. 10 are similar in shape to those in Fig. 6. Timing A and B correspond to timing A and B in Fig. 6. Timing B in Fig. 10, which is the angle at the terminal swing, is shown more clearly than that in Fig. 6. On the other hand, timing C in Fig. 11 shows the kicking power from phase 4 (terminal stance) to 5 (pre-swing) is the same as timing A in Fig. 6 and 10. However, the angle at timing D in Figure 11 is between not the foot and floor but a single limb and the vertical line to the floor. The plotted angle in Fig. 11 clearly shows a change of angle for the single limb.

As the result of this experiment, we decided that the heel was the best position to place a WD.

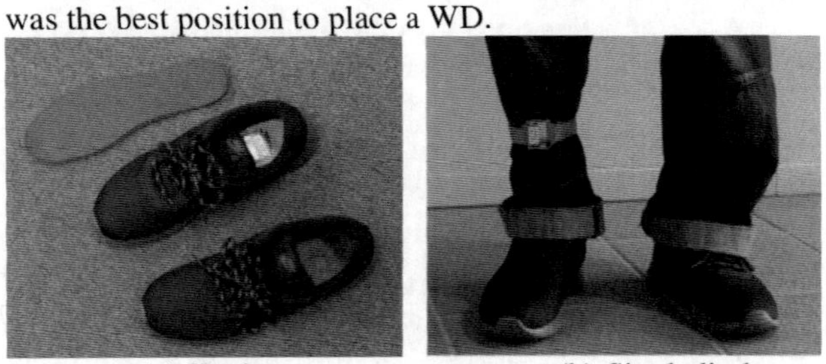
(a) Heel (b) Single limb

Figure 8. WD attaching position

Figure 9. STEVAL-WESU1 by STMicroelectronics

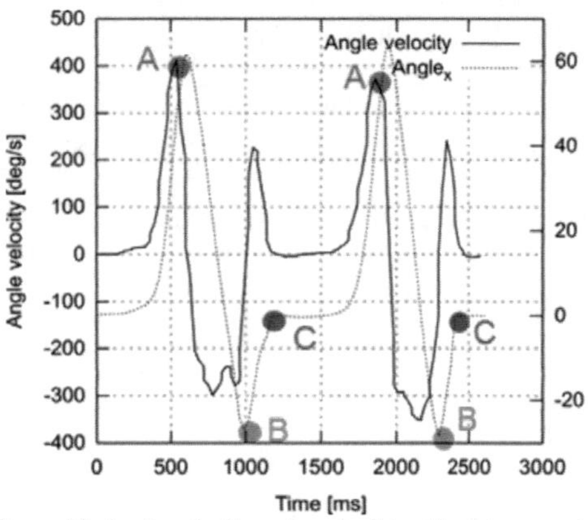

Figure 10. Angle velocity and angle data at heel in normal walk

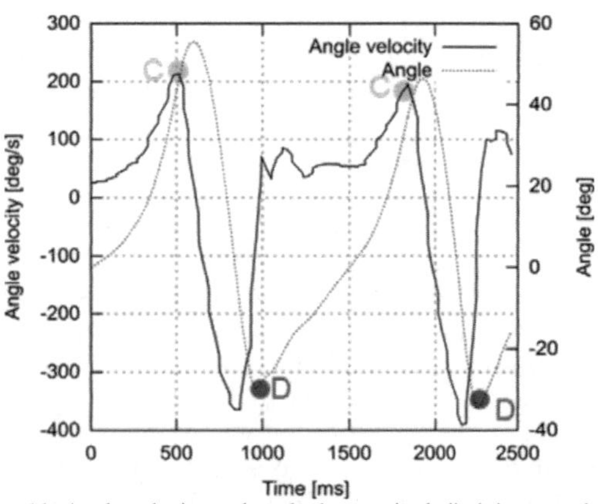

Figure 11. Angle velocity and angle data at single limb in normal walk

V. GAIT MONITORING SHOES

As described in Section IV, we determined the heel of a shoe is the best place to measure angle velocity of the foot and angle between the back part of a foot and the floor. We inserted a WD (STEVAL-WESU1 by STMicroelectronics) into soles of both shoes. And, we also developed a gait monitoring application for Android smartphone which measures and stores the angle velocity and angle as shown in Fig. 12. The upper part shows ID of WD for the right and left shoe, and the lower part shows angle velocities at A in each step for right foot, angle at B in each step for right foot, angle velocities at A in each step for left foot, and angle at B in each step for left foot. In this application, direction of angle is turned. When these graphs were measured, a participant played a stroke patient who had a one-side paralysis for the right side of the body. Therefore, most strength of angle velocity at A for right foot were smaller than that for left foot. And, most amplitude of angle at B for right foot were smaller than that for left foot.

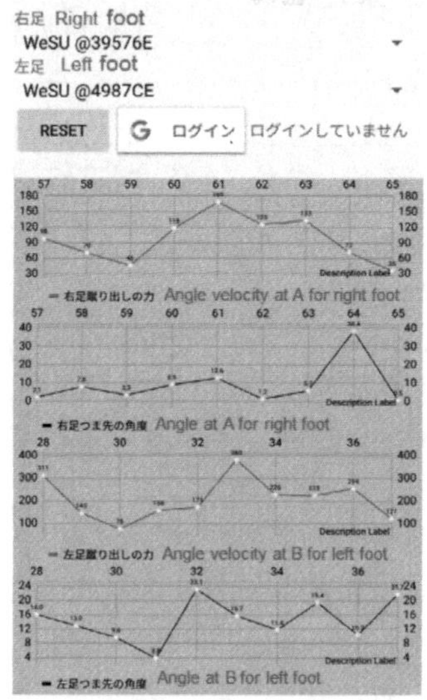

Figure 12. Gait monitoring application for Android

We experimentally monitored walking for two participants. They were unimpaired people. They walked and played the three types of trips in Fig. 2. Example measured data are shown in Fig. 13. Therefore, curves of angles in these graphs have different discontinuity to those in other graphs at the sole of a shoe touching a floor.

In (1); toe touching the floor first instead of a heel, and (2); shuffling, shapes of angle velocity resemble that of the normal walk shown in Fig. 10. However, maximum values of angle velocity and minimum angle in a cycle in Fig. 13 (1) and (2) are much smaller than those in Fig. 10. Their absolute minimum values are also much smaller than those in Fig. 10. This feature must show that when muscle strength is weaker, more trips occur.

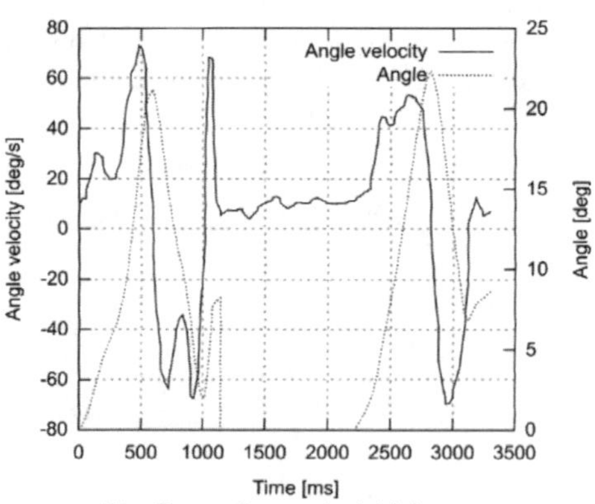

(1) Toe on floor at the initial contact

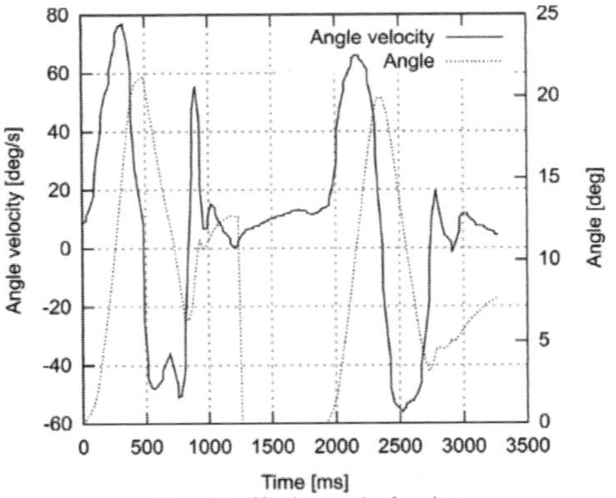

(2) Shuffle instead of swing

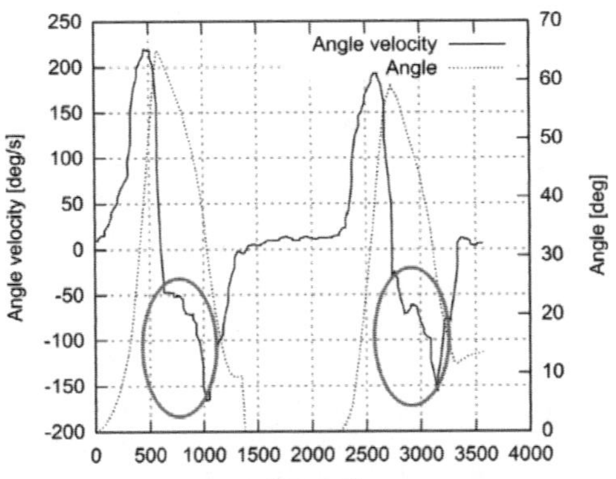

(3) Toe collision with floor at the initial swing

Figure 13. Example measured data for tripping with gait monitoring shoes

In (3), a toe dragging the floor due to it pointing down at the initial swing, the swing did not start smoothly. The red circle in Fig. 13 (3) shows this situation clearly. In the case of a normal walk, angle velocity rapidly decrease from the pre-swing to the initial swing. However, in (3), the angle velocity limply decrease on the way.

VI. CONCLUSION

We comparably analyzed walking gait cycle between normal people and stroke patients using a wearable device (WD) to detect warning signs of falls. The results of experiments for physically unimpaired students and stroke patients show that the angle velocity at the terminal stance phase and the angle between the back of foot and a floor at the terminal swing are critical parameters. We determined a shoe heel to be the best place to place a WD, inserted WDs into heels of a pair of shoes, and showed data measured using them. We plan to develop a warning system using the proposed shoes for fall prevention. More data of above critical parameters for stroke patients in daily life including tripping conditions will be measured to develop such system.

REFERENCES

[1] World Population Ageing: 1950-2050, Population Division, Department of Economic and Social Affairs, United Nations, http://www.un.org/esa/population/publications/worldageing19502050 / [retrieved: Fe, 2017].

[2] World Health Organization Regional Office for Europe, "What are the main risk factors for falls amongst older people and what are the most effective interventions to prevent these falls?," http://www.euro.who.int/en/data-and-evidence/evidence-informed-policy-making/publications/pre2009/what-are-the-main-risk-factors-for-falls-amongst-older-people-and-what-are-the-most-effective-interventions-to-prevent-these-falls [retrieved: February, 2017]

[3] F. A. Batchelor, S. M. Mackintosh, C. M. Said, and K. D. Hill, "Falls after stroke," Int J Stroke, Vol. 7, pp. 482-90, 2012.

[4] G. S. Verheyden, V. Weerdesteyn, R. M. Pickering, D. Kunkel, S. Lennon, A. C. Geurts, and A. Ashburn, "Interventions for preventing falls in people after stroke," Cochrane Database Syst Rev. (5), CD008728, 2013.

[5] J. A. Tutuarima, J. H. P. van der Meulen, R. J. de Haan, A. van Straten, and M. Limburg, "Risk Factors for Falls of Hospitalized Stroke Patients," Stroke, Vol. 28, pp. 297-301, 1997.

[6] R. Teasell, M. McRae, N. Foley, and A. Bhardwaj, "The Incidence and Consequences of Falls in Stroke Patients During Inpatient Rehabilitation: Factors Associated With High Risk," Arch Phys Med Rehabil Vol. 83, 2002.

[7] R. K. Begg, O. Tirosh, C. M. Said, W. A. Sparrow, N. Steinberg, P. Levinger, and M. P. Galea, "Gait training with real-time augmented toe-ground clearance information decreases tripping risk in older adults and a person with chronic stroke." Frontiers in Human Neuroscience, Vol. 8, Article 243, 2014.

[8] A. J. Blake, K. Morgan, M. J. Bendall, H. Dallosso, S. B. Ebrahim, T. H. Arie, P. H. Fentem, and E. J. Bassey, "Falls by elderly people at home: prevalence and associated factors," Age Ageing; Vol. 17, pp. 365-72, 1988.

[9] W. Tao, T. Liu, R. Zheng, and H. Feng, "Gait Analysis Using Wearable Sensors," Sensors, 12(2), 2012, pp. 2255-2283.

[10] Stacy J. Morris Bamberg, Ari Y. Benbasat, Donna Moxley Scarborough, David E. Krebs, and Joseph A. Paradiso, "Gait Analysis Using a Shoe-Integrated Wireless Sensor System," IEEE Transaction on Information technology in Biomedice, Vol. 12, No. 4, 2008.

[11] F. Dadashi, B. Mariani, S. Rochat, C. J. Büla, B. Santos-Eggimann, and K. Aminian, "Gait and Foot Clearance Parameters Obtained Using Shoe-Worn Inertial Sensors in a Large-Population Sample of Older Adults," Sensors 2014, Vol. 14, pp. 443-457, 2014.

[12] Mourad Benoussaad, Benoît Sijobert, Katja Mombaur, and Christine Azevedo Coste, "Robust Foot Clearance Estimation Based on the Integration of Foot-Mounted IMU Acceleration Data," Sensors 2016, 2016.

[13] William P. Berg, Helaine M. Alessio, Eugenia M. Mills, and Chen Tong, "Circumstances and consequences of falls in independent community-dwelling older adults", Age Ageing, Vol. 26, pp. 261–268, 1997.

Mood Detection and Memory Performance Evaluation with Body Sensors

Jonathan Bohbot
Department of Electrical and Computer Engineering
McGill University
Montreal, Quebec, Canada
jonathan.bohbot@mail.mcgill.ca

Zeljko Zilic
Department of Electrical and Computer Engineering
McGill University
Montreal, Quebec, Canada
zeljko.zilic@mcgill.ca

***Abstract*—This paper provides the design of a system employing an Android application connected to body sensors, which is capable of assessing the mood and memory performance of humans. The mood detection is based on the heart rate, its variability, as well as on the captured brain waves. The memory performance is evaluated based on specific brain waves observed as well. Experiments were conducted to assess the main features of the system. The mood experiment has been successful at raising the mood levels of the majority of participants when being shown stimuli composed of images and sounds. Negative or neutral mood levels could be explained by participants having other thoughts or emotions during the experiment, and by the attenuation and dampening of the body sensors' signals. The ability of participants to reach a particular mood (relaxed, engaged, and sad) more quickly in response to a conducive stimulus is related to a person's physical characteristics; for example, younger participants reach a particular mood more quickly than older participants. The memory experiment has been successful at raising the memory levels of the majority of participants when being asked to perform a modified Sternberg memory task. Results show a positive memory activity for the majority of participants, even in the presence of signal attenuation in the body sensors.**

***Keywords-Mood Detection*; *Body Sensors*; *Heart Rate Variability*; *Android Application*; *Brain Waves*.**

I. INTRODUCTION

Mood detectors can identify trends pointing to a person's mood. They can be found in modern computers, laptops, smartphones, tablets, sensors such as skin sensors, electroencephalography sensors, and voice recognition sensors. Thanks to existing technology, they can be carried everywhere and accessed anytime. Furthermore, these devices feature lower costs, high speed, low power consumption, and many other benefits for the subject, clinician or researcher.

Mood detectors can be integrated in our daily lives, such as when driving a car. The driver might feel stressed and tired when driving long distances. A car with sensors can understand the driver's emotion and feelings and prevent an accident, which could save many lives. Sensors could nest in the steering wheel and door handles to pick up electric signals from the skin. A camera mounted on the windshield could analyze facial expressions. When the driver is stressed, the car's sensors could soften the light and music, or broaden the headlight beams to compensate for the loss of vision [1].

The importance of mood detection has been increasingly recognized because it could prevent mood disorders from affecting us and harming us in our daily lives. Mood disorders include depression and bipolar disorders. Anyone can be affected from mood disorders: children, teenagers, adults, and the elderly. Stressful events in our lives can result in transitory depression, or any of several mood disorders. Symptoms include being sad, anxious, hopeless, helpless, and having low self-esteem. These symptoms can be overcome with the knowledge of one's current mood by using the above-mentioned devices and taking the right actions, such as a psychiatric consultation and/or the commencement of medication.

A. Application Features

Mood detection is a growing and rapidly developing field. Many existing devices can approximate one's current mood. Devices such as wrist sensors can provide information on stress levels when worn. They can communicate this information via the internet. For example, a sensor worn on a child's wrist might detect that the child is stressed. The stress signal is communicated via the internet and the parent can see on his/her smartphone that the child is stressed, thereby enabling the parent to take action to reduce that stress level [2].

In recent years, many applications for smartphones and tablets capable of telling a person's mood have been released. Samsung has developed a smartphone that can tell one's mood based on how the phone is used. For example, it monitors the speed at which the user is typing some text, and how much the device shakes [3]. The "mood detection and memory evaluation" Android application combines many of the features from the applications mentioned below:

- The system requires Bluetooth Low Energy in order to work.
- Mood is measured using brain and heart sensors, so it is very accurate.
- Mood data and user data is sent to the Web on a Dropbox Web Server, which is safe and secure.
- Mood data can be compared between users.
- Mood data for each user can be graphed, so it is easy for the user to compare his/her mood on a daily basis.

The paper is organized as follows: section 2 briefly discusses the system's features such as mood and memory performance algorithms and the valence/arousal model. Section 3 discusses the experimental work. Finally, Section 4 concludes on the paper work and speculates on the future of this growing field.

II. BACKGROUND

This section discusses essential background information of the work. Please note that this application is not standalone code, but is connected to body sensors communicating with it via Bluetooth Low Energy.

A. Mood and Memory Performance Algorithms

This section discusses the procedure for measuring different emotions and memory performance in real time through the use of mood and memory performance algorithms. This section also shows a pseudocode implementation of the mood and memory performance algorithms.

The recognition of different emotions and memory performance is achieved as follows:

- First, we record the resting heart rate of the participants. Any significant rise in this heart rate (when the participant is seated and relaxed) is explained by strong emotions, such as excitement, happiness, anger, and arousal [4].
- Second, we record the heart rate variability of the participant. We calculate it by using the formulas described in the next subsections. The heart rate variability is a useful feature in mood classification, since a bigger heart rate variability is an indicator of good health and a lower heart rate variability is an indicator of bad health, stress, and heart diseases.
- Third, we process the EEG data of the brain using the procedures described in the next subsection. The processed data consists of the EEG average band powers (alpha, beta, gamma, and theta). Using relations and formulas discussed in the next subsections, we obtain the arousal and valence information and we use this information along with the 2D valence/arousal model to recognize different emotions in real-time. Furthermore, memory performance evaluation follows almost the same procedure except that we look at the band powers, especially the alpha band power to evaluate retention in a short-term memory task.

1) Android Pseudocode Implementation

The pseudocode implementation of the relaxation mood (how to find the relaxation level of an individual given some parameters such as heart rate) is shown as all other mood and memory performance algorithms follow a similar implementation.

Referring to Algorithm 1, RelaxHR% measures how relaxed a person is from heart rate. A value close to 0% indicates that a person is not relaxed, and a value close to 100% indicates that a person is very relaxed. RelaxSDNN% measures how relaxed a person is from heart rate variability (ageSDNN is the ideal SDNN a person should have given his/her age). RelaxEEG% measures how relaxed a person is from EEG. Together, these values are used to find the relaxation level.

B. Data Processing

The procedure for data processing to obtain mood and memory performance levels works as follows: first, the DC offset and slow drift of the raw signal is removed. The best way to do this is with a high-pass filter with a cut-off frequency greater than 0.16Hz. The data segment is then multiplied by a tapered window function which smoothly forces the two ends of the data segment to match exactly. Step changes and start-finish differences will put fake responses into Fast Fourier Transform (FFT) data since the algorithms assume an infinitely repeating copy of the segment of data. The FFT algorithm is then executed. The FFT algorithm returns a complex set of values at each frequency increment. The magnitude of these complex numbers is squared by multiplying each complex number by its complex conjugate. Power is proportional to the square of the magnitude, so now we have the power per frequency interval. The powers are added up for each element of the frequency range of interest, so now we have the average EEG band powers.

C. Resting Heart Rate

The resting heart rate is the heart pumping the lowest amount of blood. If a person is sitting or lying, and is calm, relaxed and in good health, his/her resting heart rate should range between 60 and 100 beats per minute. A heart rate lower than 60 could be explained by a medical condition, taking drugs such as beta blockers, or being physically active and very athletic. Active people often have lower resting heart

Algorithm 1 Relaxation Mood Algorithm

```
Input: heart rate, heart rate variability (SDNN), valence, and
  arousal.
Output: relaxation level
  if (HR/HRrest*100)<100 then
    RelaxHR% = 100
  else
    if ((1-((HR / HRrest)-1))<0.0) then
      RelaxHR% = 0
    else
      RelaxHR% = ((1 - ((HR / HRRest) - 1)) * 100)
    end if
  end if
  if (age>=0&&age<=49) then
    ageSDNN = 50
  else
    ageSDNN = 40
  end if
  if (SDNN/ageSDNN*100)>100 then
    RelaxSDNN% = 100
  else
    RelaxSDNN% = ((SDNN/ageSDNN)*100)
  end if
  if (valence>0) then
    if (arousal>0.4&& arousal<=1) then
      RelaxEEG% = 100
    else if (arousal>1) then
      if ((1-(arousal-1))<0) then
        RelaxEEG% = 0
      else
        RelaxEEG% = ((1 - (arousal - 1)) * 100)
      end if
    else
      RelaxEEG% = (arousal/0.4)*100
    end if
  else
    RelaxEEG% = 0
  end if
  RL = RelaxHR%/4 + RelaxSDNN%/4 + RelaxEEG%/2
  return RL
```

rates because their heart muscle is in better condition and does not need to work as hard to maintain a steady beat [5].

D. Heart Rate Variability

Heart rate variability (HRV) is the degree of fluctuation in the length of intervals between heart beats. A bigger regularity of heart beats lowers HRV and vice versa. Regularity of heart beats is derived from a quantity of numbers equal to the time elapsed between successive heart beats. These are named R-R intervals and are measured in milliseconds (ms) [6].

HRV can be assessed in two ways: as **Time Domain Analysis**, or in the frequency domain, as **Power Spectral Density Analysis**. We will concentrate on **Time Domain Analysis** because it is more simple and straightforward.

Time domain measures are the simplest to calculate and include the mean normal-to-normal (NN) intervals during the entire recording and statistical measures of the variance between NN intervals. The most important time domain measures are the SDNN (HRV), and the RMS-SD.

- The SDNN is the standard deviation of the NN intervals, which is the square root of their variance. A variance is mathematically equivalent to the total power of spectral analysis, so it reflects all cyclic components of the variability in recorded series of NN intervals. It is inappropriate to compare SDNN values derived from the NN recordings of different lengths. A recording can last a short period of time, such as five minutes, or it can last a full 24-hour day. SDNN is measured in milliseconds [6].

- The RMS-SD is the square root of the mean squared differences of successive NN intervals. This measure estimates high-frequency variations in heart rate in short-term NN recordings that reflect an estimate of parasympathetic regulation of the heart. RMS-SD is measured in milliseconds [6].

Formulas to calculate these time domain measures are given:

Let N be the total number of heart beats. Let MRR be the mean of RR (or NN) intervals. It is calculated as follows:

$$MRR = \bar{I} = \frac{1}{N-1}\sum_{n=2}^{N} I(n) \quad (1)$$

In this formula, I(n) is the value in milliseconds of the nth NN interval. The SDNN can be expressed as:

$$SDNN = \sqrt{\frac{1}{N-1}\sum_{n=2}^{N}[I(n)-\bar{I}]^2} \quad (2)$$

Finally, the RMS-SD can be expressed as:

$$RMSSD = \sqrt{\frac{1}{N-2}\sum_{n=3}^{N}[I(n)-I(n-1)]^2} \quad (3)$$

E. Valence/Arousal Model

The 2D Valence/Arousal model is used to characterize emotions such as: happy, sad, relaxed, and angry. Emotions are characterized based on their valence and arousal values. For example, happiness is characterized by a positive valence and high arousal, anger is characterized by negative valence and high arousal, relaxation is characterized by positive valence and low arousal, and sadness is characterized by negative valence and low arousal [7].

High arousal (excitation) is characterized by a high alpha power and a low beta power (a high alpha activity and a low beta activity). The ratio of the beta to the alpha power characterizes the arousal level of a person ($Arousal = beta/alpha$). The arousal level of a person is measured using the frontal electrodes of the Emotiv Insight brain sensor. This relationship holds because beta brainwaves are associated with an alert or excited state, while alpha brain waves are associated with a relaxed state [7].

To determine the valence level, we compare the activation levels of the two cortical hemispheres. Left frontal inactivation is an indicator of a withdrawal response, which is often linked to a negative emotion. On the other hand, right frontal inactivation may be associated with an approach response, or positive emotion [7].

High alpha activity is an indication of low brain activity, and vice versa. Thus, an increase in alpha activity together with a decrease in beta activity may be associated with cortical inactivation. The frontal electrodes of the Emotiv Insight brain sensor, AF3 and AF4, are the most used positions for looking at this activity, as the frontal lobe plays a crucial role in emotion regulation and conscious experience.

We estimate the valence value in a person by computing and comparing the alpha power α and beta power β in channels AF3 and AF4, like so:

$$valence = \frac{\alpha_{AF4}}{\beta_{AF4}} - \frac{\alpha_{AF3}}{\beta_{AF3}} \tag{4}$$

III. Experimental Study

The experiment is performed to assess how effective is the system at finding the mood and memory performance levels of each participant. It presents part of the results obtained from performing the experiment and a discussion based on these results.

A. The System

The system is composed of the following components:

- External sensors, such as the Texas Instrument Sensor Tag, the Polar H7 heart rate sensor, and the Emotiv Insight brain sensor.
- The Cloud, where the Dropbox Core Api is used to store the users' data and access it anytime needed.

B. Subjects

Sixteen subjects (ages 22-80, 12 males and 4 females) performed the experiment, which consisted in evaluating their mood and memory performance levels while their heart and brain data was recorded. Informed consent was obtained from each subject prior to the study. Ethical approval was obtained from the Research Ethics Board Office of McGill University. The reference number is: 306-0116.

C. Stimulus

Both pictures and music were used to be the stimulus to elicit emotion. To represent good moods such as relaxation, happiness, engagement and arousal, pictures displaying beautiful nature scenery, people jumping out of joy, beautiful birds, plants, and animals, were selected. To represent bad moods such as sadness, anger, and stress, pictures displaying angry and wild animals, people crying, and children all alone sleeping in the streets, were selected. Three to four music pieces were chosen to represent each mood. Each lasted 30 seconds. For example, annoying alarm clock sounds were chosen to represent stress, bomb siren sounds were selected to represent anger, "Don't Worry be Happy" by Bobby McFerrin was selected to represent relaxation, "Chariots of Fire" theme song was selected to represent arousal, the theme song from the movie "Pirates of the Caribbean" was chosen to represent engagement, "Happy" by Pharrell Williams was selected to represent happiness, and "Very Sad Violin" classical music was chosen to represent sadness.

This procedure achieved stimulation by increasing and decreasing the neural activity of the brain. The cerebral cortex became synchronized at any given moment. The limbic system's cingulate gyrus, which connected actions with emotional responses, became synchronized. This synchronization happened mostly in brain parts responsible of processing sights, sounds, and emotions.

D. Testing Procedure

The experiment took a total time of 45 minutes. Data was recorded using a tablet (Samsung Galaxy Tab 4). The participant was instructed to sit and not to move his/her head for the entire duration of the experiment. The Polar H7 heart rate sensor was placed on the participant's bare chest and the Emotiv Insight brain sensor was placed on the participant's head. The data was stored on a Dropbox account. For added security, the data in the Dropbox account was encrypted using Boxcryptor. Boxcryptor features a fast and easy Dropbox encryption, a state-of-the art AES-256 encryption standard, and top security for all private and business needs. The application was launched and the participant was prompted to create an account. Creating an account requires him/her to enter a username, a password, his/her height, his/her weight, and his/her sex. The participant was prompted to take a picture of himself/herself. The air temperature, the air humidity, and the air pressure were recorded using the Texas Instrument Bluetooth Low Energy Sensor Tag. The participant was asked if he/she was relaxed or not and if not, he/she was given a short period of time (2 minutes) to relax. When relaxed, the participant's resting heart rate was recorded. Then, for each stimulus composed of pictures and sounds (each representing a particular mood such as relaxed, engaged, stressed, happy, angry, aroused, and sad) which was displayed on a laptop screen, the data was recorded on a tablet before evoking the stimulus, the stimulus was evoked, and the data was recorded on a tablet after evoking the stimulus. At the end of each stimulus, there was a relaxation period of 1 minute. The first part of the experiment took 35 minutes. In the second part of the experiment, the subject was first given a 1-minute resting period. The memory test was given immediately afterward, and the test subject's memory performance level was recorded. For the memory test, the participant was shown three sets of 4-6 letters displayed on a laptop screen. After being shown each set of letters, the participant was given a brief moment (2-5 seconds) to memorize the set of letters. The participant was then shown a letter which may or may not be part of the set of letters previously shown to him/her. The participant was asked to state whether or not this letter belongs to the set of letters. There were three sets of letters and three probes in the memory experiment. EEG data was recorded twice per set of letters: before showing the set of letters to the participant and just after showing the set of letters to the participant while he/she was busy memorizing them. The second part of the experiment took 10 minutes.

E. Results

Results were gathered from the participants' brain and heart data under the following environmental conditions: the temperature ranged from 60.28 deg/F to 67.57 deg/F, the humidity ranged from 32.25 %rH to 48.18 %rH, and the pressure ranged from 999.70 nPA to 1032.32 nPA. All participants confirmed that they were relaxed at the beginning of the experiment. All participants remained seated and did not move their heads for the entire duration of the experiment. Results for this paper contain relaxation data graphs such as the relaxation level recorded before evoking the stimulus and the relaxation level recorded after evoking the stimulus for each participant, the age, height (in meters), and weight (in kilograms) of each participant versus the difference in the relaxation level recorded after evoking the stimulus and the relaxation level recorded before evoking the stimulus. Mood and memory performance levels are in percent. A mood level of 0% suggests that the participant is not feeling at all in that particular mood, and a mood level of 100% suggests that the participant is feeling very strongly in that particular mood. A memory performance level close to 100% may be associated with strong memory activity, whereas a very low memory performance level, approaching 0%, may be associated with no memory activity.

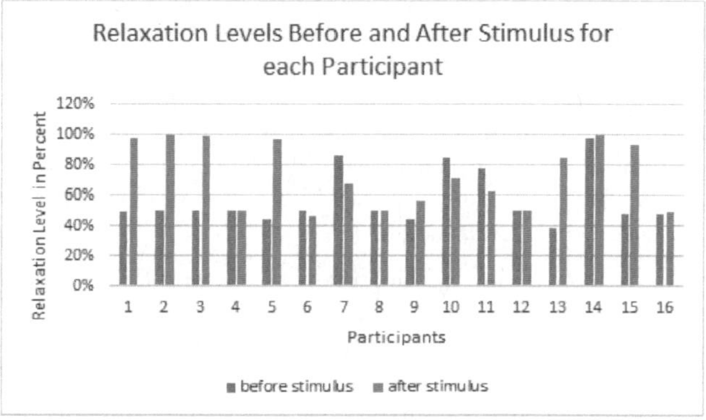

Fig. 1. Relaxation Level before and after Evoking Stimuli for each Participant

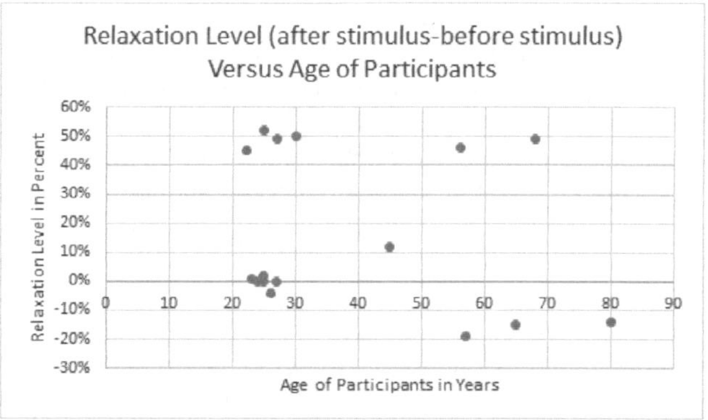

Fig. 2. Relaxation Level (after Evoking Stimuli - before Evoking Stimuli) for each Participant versus their Age

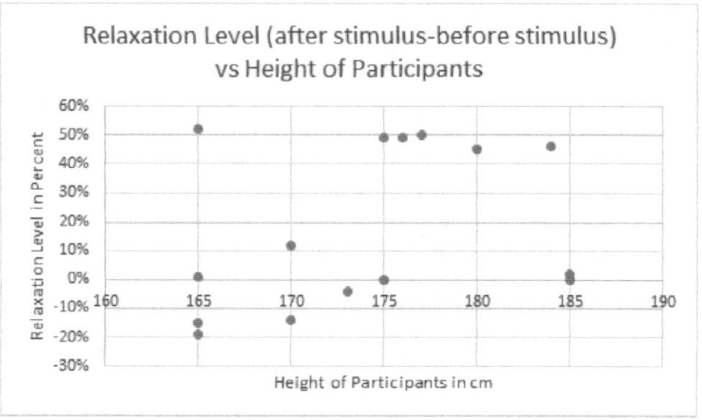

Fig. 3. Relaxation Level (after Evoking Stimuli - before Evoking Stimuli) for Each Participant versus their Height

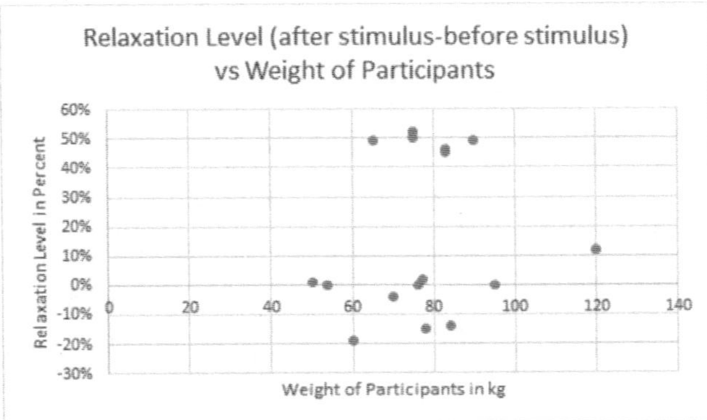

Fig. 4. Relaxation Level (after Evoking Stimuli - before Evoking Stimuli) for each Participant versus their Weight

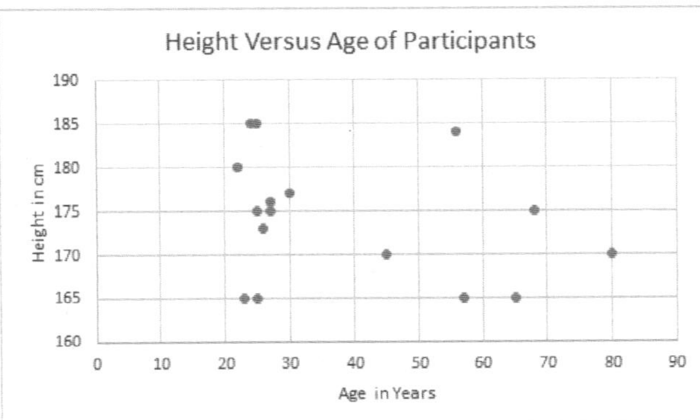

Fig. 5. Height of Participants versus their Age

F. Discussion

The experiment has been successful at raising the relaxation levels of the majority of participants. Referring to Figure 1, about 56% of all participants had a higher relaxation level after the stimulus as compared to before the stimulus. The 44% remaining participants were found to have varying reactions when exposed to such stimulus: some of the 44% may have had other thoughts or emotions during the experiment, and may have been preoccupied with other matters, thereby negating any effect of the relaxation stimulus on them. Another problem potentially accounting for the variation in results is the fact that the brain sensor's EEG data had to travel from the brain cortex to the electrodes of the brain sensor. During its travel, the signal may have been attenuated or dampened as a result of traveling through many brain regions. Thus, the quality of the signal using this method is not optimal, and results in a lower-quality EEG signal. Figure 2 showing the change in relaxation level (the relaxation level after the stimulus minus the relaxation level before the stimulus) was plotted against the age of all participants. The graph reveals that almost all participants had positive or neutral changes in relaxation levels and that most participants with the greatest positive changes in relaxation levels (higher than 40%) are the younger participants, those aged 22 to 30. The older population, those aged 40 to 80, had the most participants with negative changes in relaxation levels (3 participants aged 50-80 versus 1 participant aged 26). The results reveal that the ability of the participants to relax quickly in response to a conducive stimulus is greater for younger individuals than for older ones. The change in relaxation level was plotted against the height of all participants (Figure 3). The results reveal that the tallest individuals (heights ranging from 175 to 185 cm) had the majority of values corresponding to the highest changes (higher than 40%) in relaxation levels (5 individuals with heights ranging from 175 to 185 cm versus 1 individual with a height of 165 cm). The participants with all values corresponding to negative changes in relaxation levels are the shortest individuals (4 individuals with heights ranging from 165 to 175 cm). The results mean that taller participants are better able to relax quickly than shorter participants. The graph (Figure 4) showing the change in relaxation level against the weight of participants reveals that the majority of individuals with the greatest changes in relaxation levels are those with weights ranging from 60-90 kg. The graph also shows that the individuals from the same weight range have all values associated with negative changes in relaxation levels. Therefore, the results stating that participants with a weight range of 60-90 kg are better able to relax quickly than the other participants are inconclusive. According to the last graph (Figure 5), the majority of tall participants (those with height ranging from 175 to 185 cm) are 20 to 30 years old (7 participants with heights ranging from 175 to 185 cm versus 2 participants with the same height range who are 50 to 70 years old). The results mean that in general taller and younger participants are better able to relax quickly than shorter and older participants.

The first, second, and third trial memory performance levels before and after memorizing the set of letters for each participant show that the experiment has been successful at raising the memory performance levels of the participants. Almost all participants had higher memory performance levels after memorizing the set of letters as compared to before memorizing the set of letters which is associated with positive memory activity. The success rates of all trials are 81%. The 19% remaining participants for each trial had memory performance levels lower or equal after memorizing the set of letters as compared to before memorizing the set of letters. An explanation of the variation in results is the fact that the brain sensor's EEG data had to travel from the brain cortex to the electrodes of the brain sensor. During its travel, the signal may have been attenuated or dampened as a result of traveling through many brain regions before reaching the brain sensor's electrodes. It is very likely that the more head hair a participant had, the more attenuated or dampened the signal was when reaching the electrodes of the brain sensor.

IV. Conclusion and Future Work

This paper has presented a system capable of detecting the mood and memory performance of humans. A discussion of the mood and memory performance algorithms, data processing, resting heart rate, heart rate variability, and the valence/arousal model, was given in Section 2. Section 3 described the testing procedure and a discussion on only part of the results of the experiment was formed in order to respect the paper's length.

This application can be a first step in mental health evaluation, and, as we know, it is up to the patient to take the right decisions so that meaningful changes in health and lifestyle can be made.

To increase the accuracy of mood and memory detection in the future, we could use more external sensors such as skin sensors and infrared sensors (infrared camera to measure body temperature), all of which are connected via Bluetooth Low Energy or another protocol to a smartphone, tablet, or other electronic device. Although mood and memory evaluation from heart and brain sensors is accurate, it does not provide enough information on all body activity such as skin (perspiration), arms and legs (motion), and back (posture). The algorithms on mood and memory detection could be more refined and accurate and, when combined with the multiple external sensors, could be used to detect a bigger range of emotions. To evaluate memory, we could use more sophisticated devices such as fMRI (Functional Magnetic Resonance Imaging) and MEG (Magnetoencephalography) to get a better view of brain activity. In order to acquire such devices, experience and funding would be required.

Mood and memory evaluation devices have become very popular with the expansion and improvement of technology. The field of mood detection and memory evaluation can be expected to grow and evolve thanks to researchers and developers who seek to improve quality of life by introducing new devices and systems that help raise awareness of individuals' health and well-being. We hope that our contribution can make difference in many fields of study, especially in computer engineering, biomedicine, mental health and other medical specialties.

References

[1] CNN. Cable News Network, n.d. Web. 19 Feb. 2017. Available: http://edition.cnn.com/2014/02/04/tech/innovation/this-new-tech-can-detect-your-mood/

[2] "Sensor Detects Emotions through the Skin — MIT Technology Review." MIT Technology Review. 26 Oct. 2010. Web. 15 Jan. 2016. Available: http://www.technologyreview.com/news/421316/

[3] "New Approach for Sampling Mobile Phone Accelerometer Sensor Data for Daily Mood Assessment." New Approach for Sampling Mobile Phone Accelerometer Sensor Data for Daily Mood Assessment. Web. 16 Jan. 2016. Available: http://www.academia.edu/5041295/

[4] Feature, Katherine. "Anger Effects on Your Heart: Heart Disease, Atherosclerosis, and More." WebMD. WebMD. Web. 28 Dec. 2015. Available: http://www.webmd.com/balance/stress-management/

[5] "Resting Heart Rate." Measurement. Web. 28 Dec. 2015. Available: http://www.topendsports.com/testing/heart-rate-resting.htm.

[6] "Heart rate variability." Wikipedia. Wikimedia Foundation, 09 Feb. 2017. Web. 19 Feb. 2017. Available: https://en.wikipedia.org/wiki/Heart_rate_variability

[7] Ramirez, Rafael, and Zacharias Vamvakousis. "Detecting Emotion from EEG Signals Using the Emotive Epoc Device." Brain Informatics Lecture Notes in Computer Science: 175-84. Print.

Cumulative Benefits of Digital Health Investments in Canada

Calculating quality, access and productivity benefits on a national scale

Bobby Gheorghiu, Simon Hagens
Evaluation Services
Canada Health Infoway/Inforoute Santé du Canada
Toronto, Canada
Email: bgheorghiu@infoway-inforoute.ca
shagens@infoway-inforoute.ca

Abstract— **Capturing the benefits generated from investments in digital health is key towards demonstrating accountability to funders as well as to encouraging widespread adoption by clinicians and other health care professionals. The cumulative benefits calculation, developed by Canada Health Infoway, is a macro-level indicator trended over a period of 10 years. It represents estimated benefits accruing to various health care system stakeholders, as driven by component technologies and their associated adoption across the country. In-depth studies, validated by external experts in relevant fields, have been completed for specific technologies, such as diagnostic imaging systems, ambulatory care Electronic Medical Records (EMR), and telehealth. The financially quantifiable aspects of each study are aggregated, trended over time, and indexed to inflation. From 2007 to 2015 benefits accrued to the Canadian health care system exceeded $16B.**

Keywords-electronic health record (EHR); adoption; digital health; benefits; telehealth; electronic medical record (EMR).

I. INTRODUCTION

Efforts to build an interoperable electronic health record (iEHR) have been underway across Canada for many years. While deployment progress and rates of user adoption vary across provinces and territories, the initiative has passed its tipping point and uptake is now increasing rapidly. The potential benefits of the iEHR are substantial – improved quality of care, system efficiencies, improved access to care and use of health data to better manage the health system and facilitate ground-breaking research. An iEHR is a secure, integrated view of a person's medical records from all systems in the network; it provides a comprehensive view of a patient's medical history [1]. Typically, it integrates data from diagnostic imaging systems, laboratory information systems (LIS), as well as drug information systems (DIS), to provide a longitudinal view of a patient's clinical history. As such, it is a similar concept to that of a Health Information Exchange (HIE) [2]. Across Canada, iEHRs are at various stages of implementation and maturity and have evolved according to provincial/territorial strategies and priorities. A growing number of health care professionals in Canada can now access important information about their patients outside of their practice settings through iEHRs [3][4].

The iEHR acts as a complement to point of service systems like electronic medical records (EMR) in physician offices or clinical information systems (CIS) in hospitals. Regular measurement of adoption and maturity for these technologies has made progress easy to follow and manage. For example, in the 2015 Commonwealth Fund Survey, 73% of all family physicians reported they do use electronic records to enter and retrieve clinical notes [5].

In the interest of accountability for the public funds under its management, and optimizing the value accruing from investment of those funds, Canada Health Infoway (Infoway) has developed approaches to evaluate and systematically model the estimated value of outcomes related to select digital health solutions nationally [6]. The cumulative benefits model contains estimates for benefits generated through the use of diagnostic imaging systems, DIS, ambulatory and primary care EMRs, and telehealth.

The rest of this paper is organized as follows. Section II describes the methods used for calculating cumulative benefits. Section III describes the most recent results based on data up to the end of 2015. Section IV addresses the assumptions and limitations of the model. The conclusion, Section V closes the article.

II. METHODS

The cumulative benefits calculation is driven by individual quantitative estimates obtained from pan-Canadian studies commissioned by Infoway for each of the topics mentioned above. The concept of pan-Canadian studies to estimate national value was developed in 2006 to summarize results across diverse data domains, settings, evaluation methods and time periods. The studies aim to generate estimates, which are as comprehensive as possible, validated by experts, and reflecting available evidence. The cumulative benefits cover a subset of digital health solutions, and as such represent a portion of the value from digital health at large. They also provide an assessment of gaps and recommendations for increasing and optimizing the use and spread of technologies in order to increase value over time [6]. Estimates are expressed in Canadian dollars realized on an annual basis and base assumptions to current contexts are applied and documented. Not all outcomes represent direct financial savings, but where possible, a value is expressed financially to allow comparison of magnitudes. Where the literature does not provide sufficient evidence to quantify the current dollar value of a specific outcome, the value is omitted from quantitative modeling. In instances where a range of estimated benefits is provided, the mid-point value was used as a base estimate. The mid-point estimate for each domain-specific benefit is highlighted in Table 1 and corresponds to the year in which the study was carried out. Simplicity is a core principle of the specification of the quantitative benefits model, with most discrete value estimates derived by multiplying the magnitude of outcome observed per unit x value of outcome x extent of adoption across Canada. Adoption maturity variation (e.g., functionalities used, frequency of use, etc.) is an important driver of value in digital health deployment, so the extent of the adoption used in the model must be matched to the maturity required to achieve the magnitude outcome applied.

Adoption is determined by examining data from surveys of clinicians and patients, usage data from digital health solutions, and operational data sets collected by Infoway's partners. Specific definitions of adoption are designed to suit distinct kinds of solutions, and trended over time [6]. Adoption is measured differently for each domain or technology. This is both due to the practicality and feasibility of collecting the data and the way in which benefits were initially modelled. For example, telehealth benefits are driven by the number of clinical sessions—defined as consultations involving a clinician and a patient [7] whereas EMR benefits are driven by the number of Canadian physicians who reported using an EMR to document patient information as shown in Figure 1. EMR benefits largely accrue to physicians, and as a result, it is fitting for physician adoption to be a driving factor.

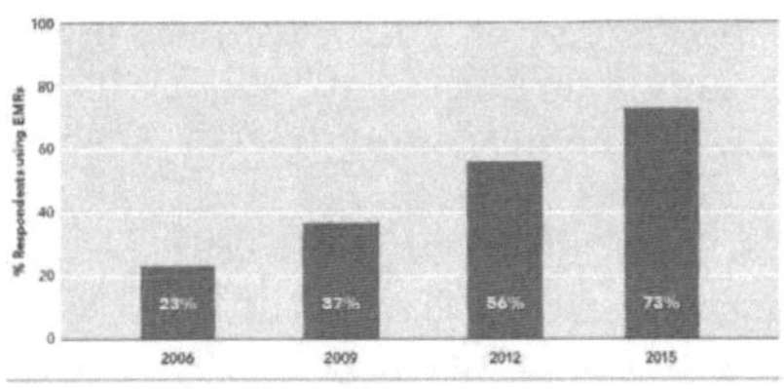

Figure 1. Family physicians in Canada reporting EMR use: Commonwealth Fund Surveys.

Conversely, telehealth benefits largely accrue to patients and the health care systems that fund patient expenses, such as travel costs. As a result, benefits are driven by patient utilization. Diagnostic imaging benefits are driven by the adoption of picture archiving and communication systems (PACS) by radiologists, physician specialists, and emergency department nurses since they are the professions who would be the most likely to depend on these systems as part of their workflow. While other clinicians and health professionals are likely to benefit, it was preferable to be conservative in the adoption estimate, in line with the organization's overall reporting principles.

Benefit estimates for previous and subsequent years are dependent on the adoption of each technology by health care professionals in each year as compared to the base years. Each year's estimate is adjusted for inflation according to the Statistics Canada Consumer Price Index for Health Care [8].

III. RESULTS

Between2007 and 2015, over $16 billion in quantifiable benefits have accrued to various parts of the Canadian health care system as shown in Figure 2. These could take the form of productivity gains for clinicians, such as those seen through the introduction of diagnostic imaging systems. Alternatively, they may represent avoided expenses for consumers, such as avoided travel by substituting an in-person visit to an urban centre with a virtual consultation via telehealth. In 2015, the estimated value to the healthcare system exceeded $2.5B.

Figure 2. Cumulative benefits of investments in digital health across

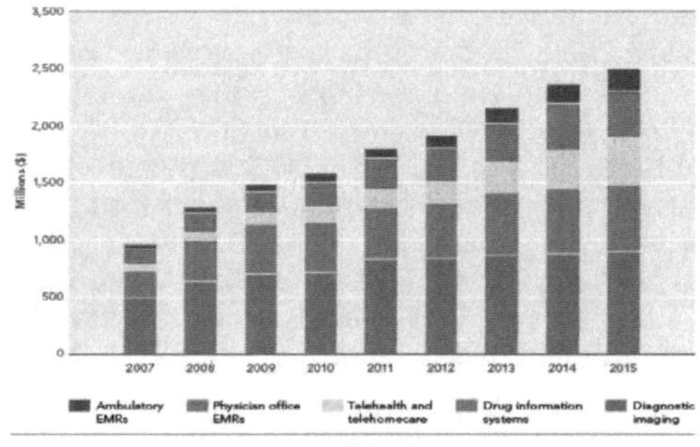

Canada.

TABLE I. YEARLY BENEFITS BY DOMAIN

	2007	2008	2009	2010	2011	2012	2013	2014	2015
DI	$511	$646	$715	$724	$841	$850	$878	$888	$908
DIS	$233	$369	$436	$441	$462	$487	$551	$579	$593
TH	$52	$70	$95	$128	$158	$187	$266	$340	$407
EMR	$131	$162	$193	$227	$270	$302	$347	$414	$419
AMB-EMR	$45	$53	$59	$71	$82	$106	$141	$167	$196
TOT.	$973	$1,300	$1,498	$1,590	$1,812	$1,932	$2,183	$2,387	$2,523

It is expected that future pan-Canadian studies, examining incremental benefits related to other components of the iEHR such as patient access to health information and/or electronic prescribing will expand the model according to the same overarching methods described above.

IV. ASSUMPTIONS AND LIMITATIONS

While it would be desirable to calculate return on investment (ROI), these benefits accrue to multiple stakeholders and costs are also borne by multiple and sometimes different stakeholders, making these calculations complicated, and best assessed on a stakeholder by stakeholder basis. As an example, productivity gains may increase access but result in additional costs to the health system since more clinical procedures and/or patient visits can be completed with the same inputs. Nonetheless, this is a useful analysis at a macroeconomic level since it can demonstrate the positive effects of investing in digital health.

The pan-Canadian studies completed to date, which drive the cumulative benefits calculation, were completed independently, over a number of years (2008 to 2015). As such, benefits calculated as part of one study, may also be reflected in another. For example, the Generation 2 Drug Information Systems study tracked patient safety benefits related to printing out prescriptions from physician EMRs—this benefit was also included in the EMR pan-Canadian study. These benefits were included only once when aggregating totals in order to eliminate double counting.

V. CONCLUSION

The cumulative benefits calculation has been and continues to be a useful tool for demonstrating the benefits of digital health investments for the purpose of accountability to funders and taxpayers. In addition, it is a useful tool to persuade clinicians and other health care professional to adopt new technologies, and to encourage partners, such as jurisdictions and health care provider organizations to continue to invest in digital health.

REFERENCES

[1] Interoperable EHRs. Canada Health Infoway. https://www.infoway-inforoute.ca/index.php/programs-services/investment-programs/interoperable-ehr. Retrieved: February 2017.

[2] Healthcare Information and Management Systems Society, HIMSS Dictionary of Healthcare Information Technology Terms, Acronyms and Organizations, Third Edition, 2013.

[3] B. Gheorghiu and S. Hagens, Measuring interoperable EHR adoption and maturity: a Canadian example, BMC Med Inform Decis Mak. Jan. 2016; doi:10.1186/s12911-016-0247-x.

[4] Canada Health Infoway. (2016). A conversation about digital health, Annual Report 2015-2016. https://www.infoway-inforoute.ca/en/component/edocman/3098-annualreport-2015-2016/view-document. Retrieved: February 2017.

[5] R. Osborn et al., "Primary Care Physicians in Ten Countries Report Challenges in Caring for Patients with Complex Health Needs," Health Affairs, vol. 34, no. 12, pp. 2104–2112, Dec. 2015.

[6] S. Hagens, J. Zelmer, C. Frazer, B. Gheorghiu, and C. Leaver, "Valuing National Effects of Digital Health Investments: An Applied Method" ITCH 2015: pp.165-169

[7] Canada Health Infoway. Telehealth Benefits and Adoption: Connecting People and Providers. 2011 https://www.infoway-inforoute.ca/en/component/edocman/333-telehealth-benefits-and-adoption-connecting-people-and-providers-full/view-document?Itemid=101. Retrieved: February 2017.

[8] Statistics Canada. Consumer Price Index, health and personal care, by province. [Online]. Available from: http://www.statcan.gc.ca/tables-tableaux/sum-som/l01/cst01/econ161a-eng.htm. Retrieved: February 2017.

Leveraging Technology to Advance Patient Engagement in a Mental Health Care Setting

Wendy Odell, Sanaz Riahi
Ontario Shores Centre for Mental Health Sciences
Toronto, Ontario, Canada
Email: odellw@ontarioshores.ca, riahis@ontarioshores.ca

***Abstract*— In 2010, Ontario Shores implemented a fully integrated electronic medical record (EMR) with the aim of enhancing patient safety and quality of care. The implementation of the EMR has created opportunities to leverage health IT to make further improvements to the quality of care we provide to our patients, including improving patient engagement through improved access to personal health information and enabling alternative care delivery models. This paper will provide an overview of the strategies Ontario Shores has employed to foster patient adoption and ongoing engagement including the launch of an EMR-integrated patient portal that enables patient and proxy access to clinical data, implementation of a mobile (mHealth) experience for patients to manage their care with actionable interventions through a secure mobile cloud technology platform, and the piloting of a virtual clinic, which will utilize e-therapy that will maximize access and utilization of evidence informed virtual treatments.**

Keywords- mental health; patient engagement; patient portal; mobile health; virtual health.

I. INTRODUCTION

The National Alliance on Mental Illness (NAMI) reports that approximately 1 in 5 adults in the U.S., 1 in 5 youth (aged 13–18) and 13% of children (aged 8–15) experience a severe mental disorder at some point during their life [1]. Yet, only about 40% of adults and 50% of children with a mental health condition received mental health services in the past year [1]. In the Canadian context, over 650,000 children in Ontario alone suffer from a mental health disorder, and suicide is cited as the second leading cause of death among youth aged 10-19 years, according to the Mental Health Commission of Canada [2].

Ontario Shores Centre for Mental Health Sciences (Ontario Shores), the first Healthcare Information and Management Systems Society (HIMSS) Davies Enterprise Award recipient in Canada and the first HIMSS EMRAM Stage 7 organization in behavioural health worldwide [3], is a recognized leader in the assessment and treatment of those living with complex and serious mental illness. Ontario Shores supports the belief that it is essential to both the health of individuals and their communities to more fully engage patients in their mental health care [4]. Studies have shown that those suffering from mental illness experience better outcomes and improved quality of life when they become self-empowered participants in their care [5]. Understanding this, Ontario Shores continuously pursues new and innovative ways to promote, measure, and improve patient engagement.

This paper will provide an overview of three strategies Ontario Shores is employing to advance patient engagement, including the launch of an EMR-integrated patient portal, implementation of an offline mobile (mHealth) experience for patients to manage their care (with actionable interventions through a secure mobile cloud technology platform), and the piloting of a virtual clinic, which will utilize e-therapy.

In section II, we outline the approaches that were employed to address patient engagement within a mental health setting. In section III, we highlight some of the challenges that we experienced with the implementation of these health IT solutions. Sections IV and V discuss some of the outcomes and conclusions that have been achieved to date as a result of implementation of these eHealth solutions including increased patient satisfaction, improved treatment/clinical outcomes, improved patient engagement and savings realized as a result of administrative efficiencies.

II. THE APPROACH/METHODOLOGY

In early 2014, Ontario Shores launched the HealthCheck Patient Portal. The implementation of the portal was aimed at enhancing patient access to their personal health information, bridging the existing gaps related to active engagement and partnership between patients, families and health care providers, evolving current practices and culture from having the provider as the "keeper of the information" to one where the provider and the patients are partners in care. It supports the paradigm shift towards patient-driven care.

Inpatients, outpatients and proxy users are able to access the portal from any device with an internet connection and can view, print or share personal health information found within the portal with other providers in the circle of care [6]. Functionality within the HealthCheck patient portal (Figure 1) includes E-views of reports, discharge summaries, allergies, and ambulatory medication, E-visits whereby secure messages can be sent to the patient's clinician, and E-requests for prescription renewals (Figure 2) [6].

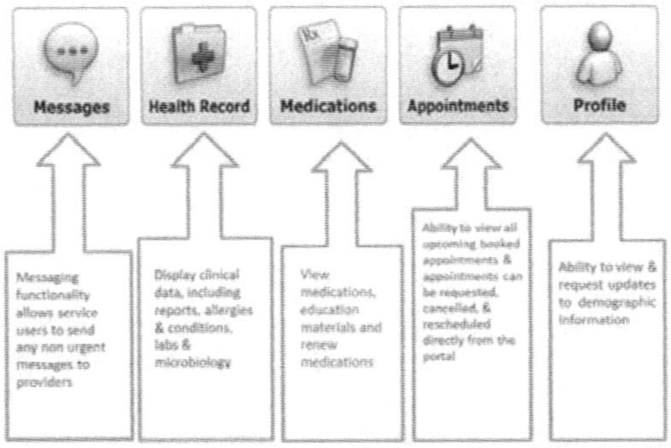

Figure 1. Healthcheck patient portal main page tabs and functionality

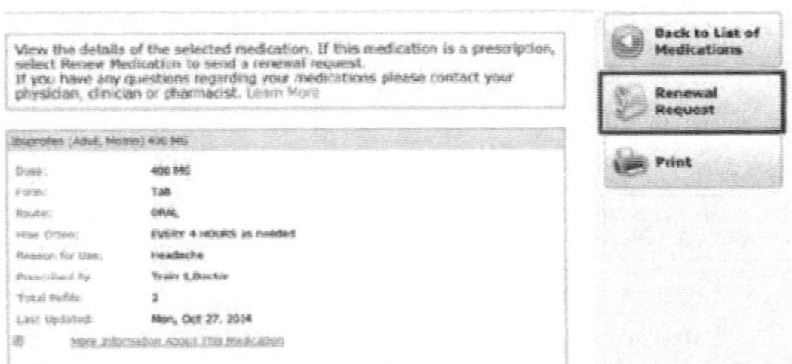

Figure 2. View of e-request prescription renewal in Healthcheck Patient Portal

Leveraging this engagement solution, Ontario Shores has continued to drive patient engagement forward by partnering with a technology based company that offers secure mobile and cloud technology platforms that will integrate with the EMR. It will enable a mobile remote-patient-monitoring health application that delivers personalized interventions to individuals in support of their care plans, thus supporting patients through interactive strategies. Not only will this solution enable patients to be actively engaged in their care through the completion of various behaviour assessment scales, meal assessments, and thought/behaviour trackers, it will also support predictive analytics capabilities, enabling clinicians to more proactively monitor patients and respond to changes in behaviours and assist with identifying patterns and triggers that could prompt early clinical interventions, the development of effective coping strategies, and possibly avoid hospital admissions. Figure 3 illustrates Ontario Shores' mobile health technology solution.

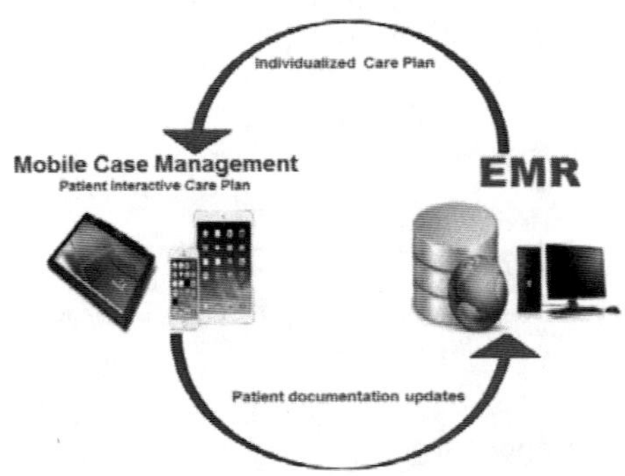

Figure 3. Ontario shores' mobile health technology patient engagement solution

Ontario Shores is also designing and implementing a virtual clinic as an adjunct to the current ambulatory Traumatic Stress Clinic following a stepped-care approach. The virtual clinic is aimed at bridging the long wait times for this clinic with the goal of maximizing patient access to care, utilizing evidence-informed virtual treatments, while more efficiently utilizing clinical resources. The virtual clinic has the potential to target improved population health, to optimize the patient experience and increase value for money.

III. CHALLENGES/BARRIERS

Transparency is an essential component for a healthcare organization interested in achieving improved patient engagement. One of the many benefits of patient portals is that they promote transparency. With this increased transparency, however, effectively integrating change management strategies to support clinicians with the transition to the use of a patient portal is paramount to how clinicians perceive the implementation of the portal. Recall that this is a paradigm shift for both patients and clinicians, and patient access to records can be a sensitivity for clinicians who feel that their practice may be called into question. At Ontario Shores, clinicians required education and support with shifting their documentation practices to ensure that it supported the transition from documentation that focused solely on sharing information with the treatment team and/or community partners to one which is aimed at achieving a partnership in care with patients via transparent and meaningful personal health information.

In addition, limitations related to functionality are also at play, which can impact adoption by patients and endorsement by clinicians. Currently, the portal system does not provide access to the full EMR (progress notes are currently not viewable, for example) and therefore portal users may still be required to submit a request for information to view components of their EMR that are not currently viewable.

Additional barriers and risks include patient access and adoption of technology (internet access, mobile tablets or phones, Personal Computers (PCs), language barrier, visual and auditory disabilities). Patients with complex health needs may also experience difficulty in learning or using the mobile applications. These risks can be managed by providing training and guidance on content as well as ensuring a comprehensive accessibility plan.

It is also important to note the role that device selection plays in adoption. Planning for the availability of devices for the implementation is integral to the process. Engaging end users, both patients and clinicians, in device selection is also essential to successful adoption. Devices that are cumbersome to use will have a negative impact on patient adoption and clinician endorsement.

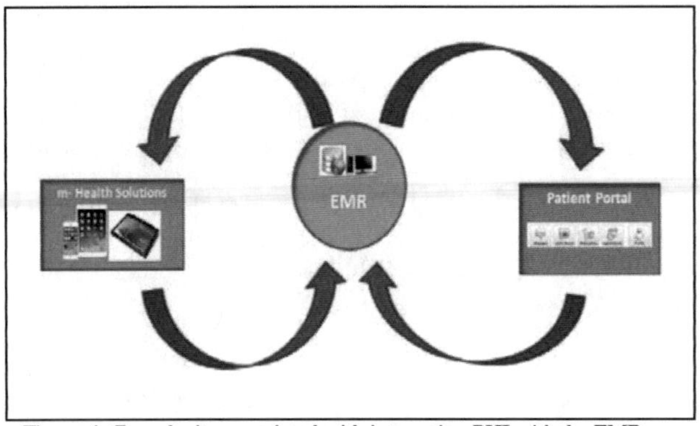

Figure 4. Complexity associated with integrating PHI with the EMR and mHealth solutions

Finally, overcoming interoperability challenges is central to a successful implementation. The complexity of integrating patient health information with the EMR and mHealth solutions must be well understood. Figure 4 demonstrates the flow of information that needs to occur in order to interface the eHealth solutions with the EMR.

IV. OUTCOMES ACHIEVED TO DATE

Following implementation of the patient portal, a rigorous benefits evaluation research study was conducted to assess the impact of the portal on patient engagement, satisfaction, and health outcomes. The results of this evaluation provide implications for enabling active patient participation in their care through transparent access to their Personal Health Information (PHI). They also point to increased patient activation, improved organizational productivity, and greater administrative efficiency.

Using the Mental Health Recovery Measure (MHRM), a robust self-reporting instrument designed to comprehensively assess the recovery process for patients with serious mental illness, along with other system-generated reports to evaluate the impact of the patient portal implementation, Ontario Shores saw a nearly 10% decrease in missed appointments for portal users, an 86% reduction in portal users' release of information requests (Figure 5), and an administrative time savings of between 10.5 and 40 hours per portal user. In addition, statistically significant improvements were seen in portal users in six domains of the MHRM scores following six months of portal access including basic functioning, overall well-being, spirituality and advocacy/ enrichment (Figure 6).

2014 Total ROI Requests Made	2015 Total ROI Requests Made	Total ROI Percentage Decrease
23	3	86%

Figure 5. Portal users total frequency of requests of information (ROI)

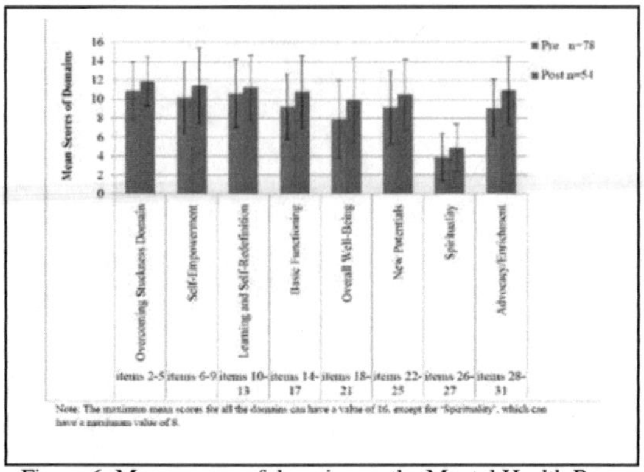

Figure 6. Mean scores of domains on the Mental Health Recovery Measure (MHRM) at pre and post-enrolment to the Healthcheck Patient Portal

The implementation of our mobile technology solution, currently underway, has the potential to advance and support areas of patient engagement that do not currently exist at Ontario Shores, such as tailoring engagement to each patient's unique learning style, continued care and therapy using asynchronous methods that do not rely on traditional face-to-face interactions, creating a feedback loop to clinicians that enables predictive monitoring, meeting the unique needs related to patient instructions, discharge education needs, and supporting a seamless transfer of care.

Expected outcomes include improved treatment and medication adherence, decreased missed appointments as a result of appointment reminders, and improvement in the quality of care plans that support an increased understanding of the triggers that lead to behavioural or emotional difficulty. Additionally, we expect to see improvement in Recovery Assessment Scale-Revised (RAS-R) Scores, which measures a patient's sense of recovery in five domains- (1)personal self-confidence and hope, (2) willingness to ask for help, (3) goal and success orientation, (4) reliance on others, and (5) life view beyond their symptoms [7]. The tool is used to facilitate collaborative, recovery-oriented practice and measure recovery-focused outcomes over time that will enable clinicians to make data-driven decisions to support care planning and assist patients in formulating recovery-oriented goals [7]. Figure 7 shows the percentage of patients that have had an RAS-R score increase by greater than 5 pre-implementation on the pilot units/programs.

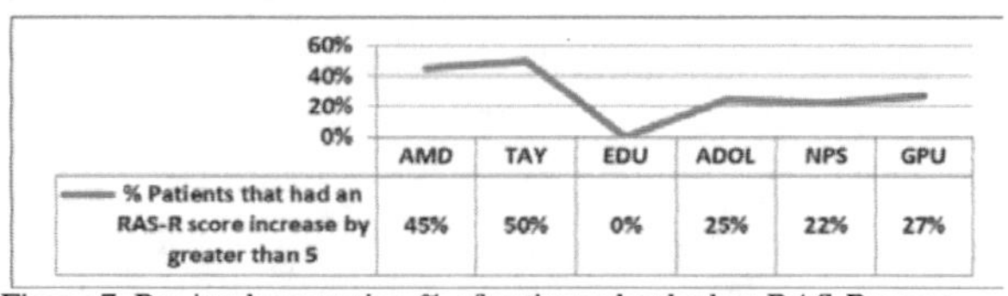

Figure 7. Pre-implementation % of patients that had an RAS-R score increase by greater than 5

Likewise, our virtual clinic is also in the implementation phase and the evaluation will form the basis for recommendations to expand virtual care delivery. Ultimately, the aim of this model is to address the growing number of patients on the waitlist (currently 247), which will improve patient flow and access to services as well as decrease the average wait time for treatment, currently 391 days. Outcome indicators that will be tracked include a clinically significant change in Post-Traumatic Stress Disorder (PTSD) Severity, which will be assessed using the 20-item PTSD Checklist (PCL-5) scale which will assess the 20 DSM-V symptoms of PTSD, a change in depression severity which will be assessed using the Patient Health Questionnaire 9 (PHQ-9), and in-person services avoided (i.e., increased efficiency) based on the percentage of patients that will not require formal, in-person services because they experienced improvement by using the virtual clinic.

V. CONCLUSION

Ontario Shores has made substantive strides towards improving patient satisfaction. One early adopter of the portal indicated "[j]ust having my own access has given me freedom as a patient." This type of qualitative feedback is indicative of patients feeling increased autonomy as a result of having access to their PHI. Additionally, users indicated that "[t]his system is very helpful for appointment reminders", which is supported by the downward trend in overall missed appointment percentages post-portal implementation. Missed appointments among portal users decreased by 9% in comparison to non-users who decreased by 6%. In 2015, users accounted for 26% of 4948 total missed appointments compared to non-users who accounted for 74%. In addition, users accounted for 958.5 hours planned for appointments that were not fulfilled or 0.5 full-time equivalent (FTE), while non-users accounted for 2752.5 hours planned for appointments or 1.4 FTE.

In terms of the impacts that these strategies have had on improved patient outcomes, statistically significant improvements were seen in portal users within the following six domains of the MHRM scores following six months of portal access: overcoming stuckness, basic functioning, overall well-being, new potentials, spirituality, and advocacy/enrichment. These results add to the body of knowledge and literature, as there are currently very limited studies focusing on the impacts of patient portals on users among the mental health population.

Given the vital role that predictive analytics can play in the area of mental health, the collection of application usage data, self-reported assessments and behavior tracking through the mHealth app, clinicians can track patient behaviors, identify triggers and can proactively implement preventative interventions to avoid hospital admissions and emergency room visits.

Additionally, patients can be directed and prompted with adaptive behaviours to improve self-care using the app. Patients can participate in manualized therapies at their convenience through their mobile device anytime and anywhere. By incorporating this type of technology into patient care, we are realizing goals around enhancing access to quality mental health care.

In terms of additional value for money post portal implementation, the organization saw a decrease of 86% in the frequency of ROIs compared to non-users who saw a decrease of 57%. Users contributed to an organizational time savings from managing requests of between 10.5-40.0 hours.

ACKNOWLEDGMENT

The authors would like to thank Canada Health Infoway, who provided financial support and the framework and guidance for the benefits evaluation for the HealthCheck patient portal implementation.

The authors would also like to acknowledge Ontario Centres of Excellence who are providing funding for the mHealth solution project through the AdvancingHealth program.

REFERENCES

[1] "National Alliance on Mental Illness," 2016, URL: http://www.nami.org/Learn-More/Mental-Health-By-the-Numbers [accessed 2017-02-27].

[2] "Working with Children and Youth with Complex Mental Health Needs. Canadian Mental Health Associations," 2015, URL: http://www.complexneeds.ca/about-the-project/ [accessed 2017-02-27].

[3] HIMSS Transforming Healthcare through IT. 2015. http://www.himss.org/news/ontario-shores-centre-mental-health-sciences-selected-2015-himss-enterprise-davies-award-recipient [accessed 2017-01-09].

[4] S. Greenfield, S. Kaplan, and JE. Ware Jr, "Expanding patient involvement in care: Effects on patient outcomes," Ann Intern Med, vol. 102, pp. 520-8, 1985, doi: 10.7326/0003-4819-102-4-520.

[5] M. Farkas, "The vision of recovery today: what it is and what it means for services," World Psychiatry, vol. 6, pp.68–74, 2007.

[6] S. Kipping, M. I. Stuckey, A. Hernandez, T. Nguyen, and S. Riahi, "A Web-Based Patient Portal for Mental Health Care: Benefits Evaluation," J Med Internet Res, vol. 18, November 2016. URL: https://www.jmir.org/2016/11/e294 [accessed 2017-02-27].

[7] P. Burgess, J. Pirkis, T. Coombs, and A. Rosen, "Australian Mental Health Outcomes and Classification Network Sharing Information to Improve Outcomes- An Australian Government funded initiative Review of Recovery Measures, version 1.01," February 2010, URL: http://www.mentalhealth.va.gov/communityproviders/docs/review_recovery_measures.pdf [accessed 2017-02-27].

An Ultra-Band Study of Pulse Rate Variability for Homecare by Using Instantaneous Pulse Rate Variability

Po-Hsun Huang
Institute of Computer Science and Engineering,
National Chiao Tung University
Hsinchu, Taiwan R.O.C.
e-mail: sdk58927@gmail.com

Chia-Chi Chang
A-Neuron Electronic Corporation
Institute of Biomedical Engineering,
National Chiao Tung University
Hsinchu, Taiwan R.O.C.
e-mail: ccchang.cs@gmail.com

Tzu-Chien Hsiao
Department of Computer Science, Institute of Biomedical Engineering, Biomedical Electronics Translational Research Center
National Chiao Tung University
Hsinchu, Taiwan R.O.C.
e-mail: labview@cs.nctu.edu.tw

Abstract—**Most of human body regulation is controlled by the autonomic nervous system (ANS). The activities of ANS influence the human health, therefore, how to observe the proper activity of ANS in time is important. Heart rate variability (HRV) is generally a common and widely non-invasive used method for indicating the activities of ANS. However, electrocardiography (ECG) used for HRV analysis is not suitable for homecare. Pulse rate variability (PRV) of photoplethysmography (PPG) is a surrogate method for HRV of ECG. Furthermore, PPG is easier and more convenient to use at home. Whether using HRV or PRV, there still remains some limitations because of the timescale which causes the frequency domain analysis restricted to only 0.5 Hz. For breaking the limitations, a novel method instantaneous pulse rate variability (iPRV) was proposed. The previous study had shown the difference between fever patients and normal people by using iPRV analysis. Nevertheless, it did not compare the result of iPRV to the conventional method. The aim of this study is to examine that iPRV can not only indicate the activities of ANS as PRV but provide more information. The result of this study shows the conventional indexes have no significant difference between PRV and iPRV. But the modified indexes including the very high frequency band (VHF) computed by PRV are different to the one computed by iPRV. The power of VHF is much greater in iPRV than PRV. This band as an indicator may contain more information about the body regulation. In this study, using iPRV as a method in homecare has more efficiency.**

Keywords-Pulse rate variability (PRV); Empirical mode decomposition (EMD); Instantaneous PRV (iPRV).

I. INTRODUCTION

There are various mechanisms for regulating the human body. Many of them are dominated by the autonomic nervous system (ANS). The ANS contains sympathetic (SNS) and parasympathetic (PNS). The balance between SNS and PNS is important to maintain health of human. It influences the health when the activities of ANS are not stable [1].

Heart rate variability (HRV) is generally a common and widely non-invasive used method for indicating the activities of ANS [2]. Electrocardiography (ECG) signal is acquired to calculate the beat-to-beat interval (RRI) for HRV analysis (Fig 1(a)). But, ECG measurement is too inconvenient to use in the home. It is not suitable for homecare.

Compared to ECG, photoplethysmography (PPG) is easier and more convenient to use in the home [3]. It is called pulse rate variability (PRV) if applying the PPG signal instead of ECG to calculate the peak-to-peak interval (PPI) for HRV analysis (Fig 1(b)). Though the ECG signal and PPG signal have the difference because of the pulse transition time (PTT) between RRI and PPI, many studies have determined that PRV can act as a surrogate of HRV during some conditions [4].

Nevertheless, the timescale of RRI and PPI have a limitation. It causes the frequency band of HRV or PRV spectrum analysis to be restricted to about only 0.5 Hz [5].

For breaking the limitation, instantaneous pulse rate variability (iPRV) was proposed [6]. iPRV adopted PRV technique and applied the frequency range extension method based on Hilbert-Huang transform (HHT) [7]. HHT can deal with the non-stationary and nonlinear data by the pre-process, called empirical mode decomposition (EMD). However, the intermittency phenomenon, which called mode mixing

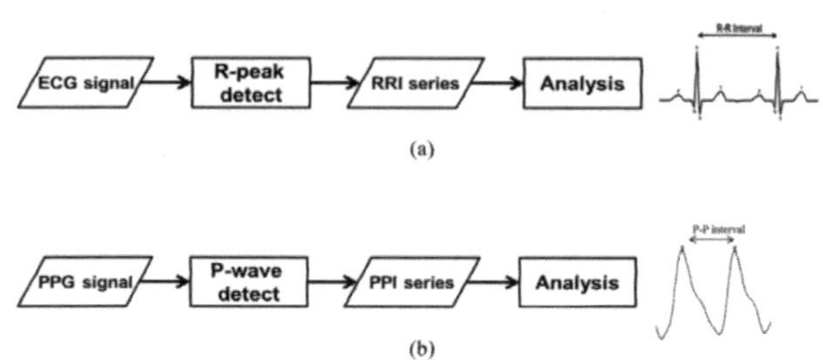

Figure 1. The procedure of (a) HRV and RRI, (b) PRV and PPI.

problem is involved in EMD. The noise-assisted method called ensemble EMD (EEMD) was proposed to solve this problem [8].

The previous study had compared the difference between the fever patients and normal people by iPRV spectrum analysis [9]. The aim of this study is to compare the iPRV spectrum result of the previous study [9] to PRV result to examine iPRV is useful and reliable for observing the activities of ANS. Besides this, the iPRV has more information in embedded in higher frequency band than PRV.

II. MATERIA AND METHOD

A. *Subjects and data collection*

The experiment was carried out in Yo-Yo Clinic, Kaohsiung, Taiwan. The body temperature was measured by ear thermometer (Radiant TH889, Radiant Innovation Inc.) and the signal was acquired by PPG (Nonin 8500, Nonin Medical Inc.) with 200 Hz sampling rate.

30 subjects were recruited in this study. 15 subjects whose body temperature are higher than 37.9 °C serve as fever group, others serve as normal group. All subjects were measured the body temperature before acquiring the PPG signal. The subjects were required to rest quietly in supine position when the PPG signal acquiring for 10 minutes. This experiment was approved by the institutional review board of the National Chiao Tung University. Informed consent was obtained from all subjects before the experiment.

B. *PRV procedure*

The procedure PRV is shown in Fig 2. First, finding out the peaks in each pulse cycle. Second, calculating the PPI series by the interval between each peak and interpolating to 200 samples per second. Last, transforming the PPI series to power spectrum by fast Fourier transform (FFT) and calculating the power of each band.

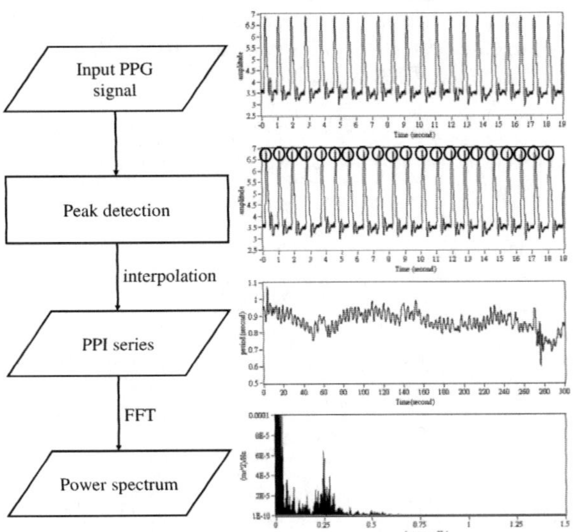

Figure 2. An example procedure of PRV.

C. *iPRV procedure*

There are many steps of the iPRV procedure [6]. First, it uses EMD to decompose the PPG signal to many intrinsic mode functions (IMFs) [7]. The algorithm of EMD is shown in Fig 3. However, because of the mode mixing problem involve in the EMD, the EEMD was used in this study. The procedure of EEMD is shown in Fig 4. [8]. After decomposition, one of the IMFs which was sinusoid-like is considered as the heartbeats component. This IMF is used for calculating the instantaneous frequency (IF) by normalized direct quadrature (NDQ) [10]. The procedure of NDQ is shown in Fig 5. The iPRV uses the inversion of IF called iPeriod to calculate the power spectrum by FFT. The whole procedure of iPRV is shown in Fig 6. The spectral analysis programs in this study

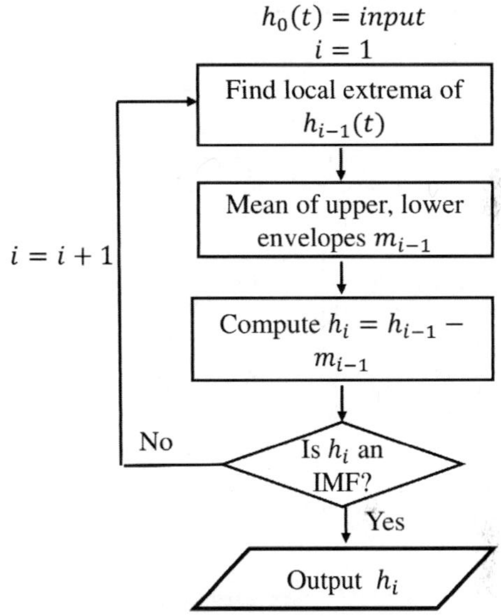

Figure 3. The algorithm of EMD.

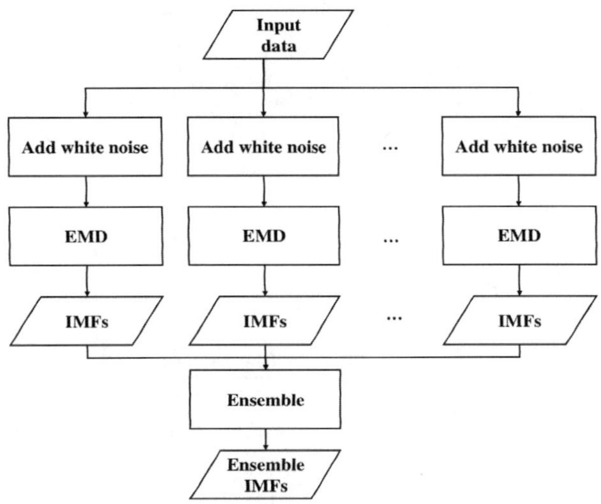

Figure 4. The algorithm of EEMD

were developed by using a commercial software platform (LabVIEW version 2013, National Instruments Corp., Austin, USA).

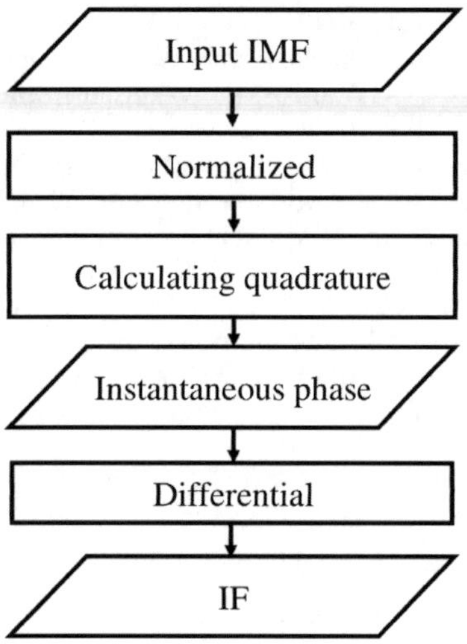

Figure 5. Calculate IF by NDQ.

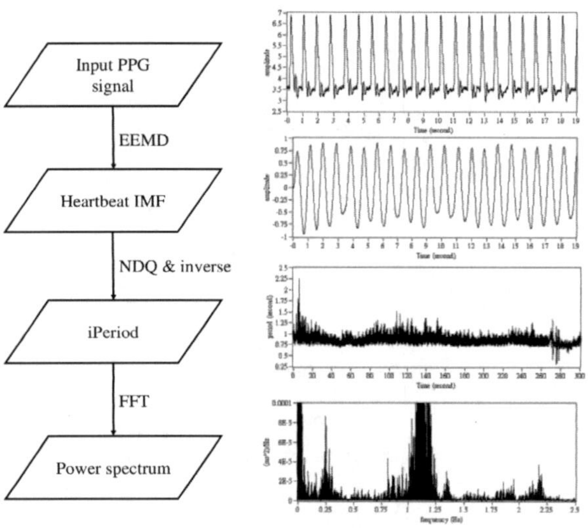

Figure 6. The procedure of iPRV.

D. *Iindex calculating*

The standard of bandwidth of low frequency band (LF) is between 0.04 Hz and 0.15 Hz and high frequency band (HF) is between 0.15 Hz and 0.4 Hz [11]. The conventional indexes of normalized power (nLF and nHF) and LF-HF ratio are calculated as follows.

$$nLF = 100\% * (LF / (LF + HF)), \tag{1}$$

$$nHF = 100\% * (HF / (LF + HF)), \text{ and} \tag{2}$$

$$\text{LF-HF ratio} = 100\% * (LF / HF). \tag{3}$$

For calculating including very high frequency band (VHF), we assume the bandwidth of VHF is from 0.4 Hz to 0.5 Hz in PRV and 0.4 Hz to 0.9 Hz in iPRV. The modified indexes are calculated as follows.

$$nLF^{a} = 100\% * (LF / (LF + HF + VHF)), \tag{4}$$

$$nHF^{a} = 100\% * (HF / (LF + HF + VHF)), \text{ and} \tag{5}$$

$$nVHF = 100\% * (VHF / (LF + HF + VHF)). \tag{6}$$

Each abbreviation, LF, HF and VHF in equations presents the power of each frequency band.

E. *Statistic analysis*

Homogeneity test was used to test the distribution of each index calculated by PRV and iPRV is homogeneity or not before independent t-test. Independent t-test was used to compare the significant difference between each index calculated by PRV and iPRV. *P* value of <0.05 was considered significant. All statistical analysis was performed by using commercial statistics software (IBM SPSS statistics, version 22.0.0.0, IBM corp., New York, USA).

III. RESULT

The comparison result between PRV and iPRV is shown in Table 1.

TABLE 1. THE COMPARISION BETWEEN PRV AND IPRV

group	**Normal group**		**Fever group**	
method	***PRV***	***iPRV***	***PRV***	***iPRV***
nLF	49.6±16.6*	46.5±16.4*	67.2±12.4	62.5±12.8
nHF	50.4±16.6*	53.5±16.4*	32.8±12.4	37.5±12.8
LF-HF ratio	128.8±112.6*	109.9±86.9*	263.6±187.0	207.5±136.2
nLF^{a}	42.8±15.8*#	26.4±10.9	56.7±11.9#	32.5±11.7
nHF^{a}	43.9±16.6*#	30.5±11.5*	27.5±10.1#	18.8± 7.4
nVHF	13.2±10.2#	43.1±10.4	15.8±6.8#	48.7±13.6

1 * means p-value <0.05 compared with Fever group, # means p-value<0.05 compared with iPRV

There is no significant difference in nLF and nHF between PRV and iPRV. If calculating including VHF, each index in PRV is significantly different to the one in iPRV whether in normal group or fever group. However, each index whether computed including VHF or not, they have the same trend in PRV and iPRV. For example, comparing to normal group, nLF increased in fever group whether calculated by PRV or iPRV. Nevertheless, the value of nVHF is much larger in iPRV than PRV, we discuss it in the next section.

IV. Discussion

The conventional indexes, nLF, nHF and LF-HF ratio, calculating by HRV usually present the activities of ANS [2][11]. The PRV can be used as an alternative measurement of the HRV [4]. So, nLF, nHF and LF-HF ratio are useful for indicating the activities of ANS by PRV. Besides this, the result of this study also shows the value of nLF, nHF and LF-HF ratio calculated by iPRV are no significant difference to PRV. Therefore, nLF, nHF and LF-HF ratio calculated by iPRV can also indicate the activities of ANS. For modified indexes in this study, because of the narrow bandwidth of VHF in PRV, nLF^a and nHF^a are similar to the conventional indexes. Otherwise, the power of VHF is much larger in iPRV. VHF band may contain more information about not only activity of ANS but more body regulation.

Some studies have assumed that VHF has the possible meaning of cardiac output or peripheral circulation [6][12][13]. Another study also shows the VHF band has potential to evaluate fluid responsiveness [14]. In addition, the previous study had compared the difference between normal people and fever patient and discussed the relationship between some thermoregulation and each index [9]. However, it still needs more experiments to examine the meaning of VHF.

V. Conclusion

This study has shown the reliability of the conventional indexes in iPRV. Moreover, the ultra-band, VHF, computed by iPRV contains more information about the body regulation for monitor. iPRV uses the convenient and easy measurement for signal acquiring, but provide more information than PRV. iPRV has a potential to be a better monitor of health status for homecare.

Acknowledgment

This work was fully supported by the Taiwan Ministry of Science and Technology under grant numbers MOST-105-2221-E-009-159, MOST 105-2634-E-009-003, and in part by 'Aim for the Top University Plan' of the National Chiao Tung University and Ministry of Education, Taiwan, R.O.C.

References

[1] L. K. McCorry, "Physiology of the autonomic nervous system," American journal of pharmaceutical education, Vol. 71(4):78, Aug. 2007, pp. 1-11.

[2] S. Akselrod, et al., "Power spectrum analysis of heart rate fluctuation: a quantitative probe of beat-to-beat cardiovascular control," Science, Vol. 213, 1981, pp. 220-222.

[3] J. Allen, "Photoplethysmography and its application in clinical physiologicalmeasurement," Physiological measurement, Vol 28(3), 2007, pp1-39

[4] E. Gil, et al., "Photoplethysmography pulse rate variability as a surrogate measurement of heart rate variability during non-stationary conditions," Physiological Measurement, Vol. 31, Aug. 2010, pp. 1271-1290

[5] G. G. Berntson, et al., "Heart rate variability: origins, methods, and interpretive caveats," Psychophysiology, Vol. 34, 1997, pp. 623–648.

[6] C. C. Chang, T. C. Hsiao, and H. Y. Hsu, "Frequency range extension of spectral analysis of pulse rate variability based on Hilbert–Huang transform," Medical & Biological Engineering & Computing, Vol. 52, 2014, pp. 343-351.

[7] N. E. Huang, et al., "The empirical mode decomposition and the Hilbert spectrum for nonlinear and non-stationary time series analysis," Proceedings of the Royal Society of London. Series A: Mathematical, Physical and Engineering Sciences, Vol. 454, Mar. 1998, pp.903-995

[8] Z. Wu and N. E. Huang, "Ensemble empirical mode decomposition: a noise-assisted data analysis method," Advances in Adaptive Data Analysis, Vol. 1, Jan. 2009, pp. 1-41.

[9] P. H. Huang, C. C. Chang, and T. C. Hsiao, "The analysis of youngster with fever by using instantaneous pulse rate variability," The 8th International Conference on eHealth, Telemedicine and Social Medicine, (eTELEMED 2016) IARIA, 2016, pp. 63-67, ISBN: 978-1-61208-470-1.

[10] N. E. Huang, et al., "On instantaneous frequency," Advances in Adaptive Data Analysis, Vol. 1, Apr. 2009, pp. 177–229.

[11] M. Malik, "Heart rate variability: standards of measurement, physiological interpretation, and clinical use," European Heart Journal, Vol 17, 1996, pp. 354-381

[12] C. C. Chang, H. Y. Hsu, and T. C. Hsiao, "The interpretation of very high frequency band of instantaneous pulse rate variability during paced respiration," Biomedical Engineering Online, Vol. 13, Apr. 2014, pp. 46-56

[13] P. C. Lin, C. C. Chang, T. C. Hsiao and H. Y. Hsu, "The circulation assessment of daily e-health by using instantaneous pulse rate variability during nonstationary conditions," The 7th International Conference on eHealth, Telemedicine and Social Medicine, (eTELEMED 2015) IARIA, Feb, 2015, pp. 209–212, ISBN: 978-1-61208-384-1.

[14] P. C. Lin, C. C. Chang, T. C. Hsiao and H. Y. Hsu, "A short-term assessment of cardiac output by using instantaneous pulse rate variability ,"The 8th International Conference on eHealth, Telemedicine and Social Medicine, (eTELEMED 2016) IARIA, 2016, pp. 110-115, ISBN: 978-1-61208-470-1 115

Application of Holo-Hilbert Spectral Analysis on Human Breathing Movement with Isovolume Maneuver

Jia-Hua Lee
Institute of Biomedical Engineering
National Chiao Tung University
Hsinchu, Taiwan R.O.C.
e-mail: jessie1992a@gmail.com

Ya-Chen Chen
Institute of Computer Science and Engineering
National Chiao Tung University
Hsinchu, Taiwan R.O.C.
e-mail: milktea0623@gmail.com

Tzu-Chien Hsiao*
Department of Computer Science, Institute of Biomedical Engineering, Biomedical Electronics Translational Research Center
National Chiao Tung University
Hsinchu, Taiwan R.O.C.
e-mail: labview@cs.nctu.edu.tw

***Abstract*—Breathing is one of essential processes for human to inhale and exhale air for gas translation. People modulate respiratory frequency (fast or slow) and depth (deep or shallow) consciously or unconsciously. Different breathing patterns perform distinct breathing movement and physiological mechanism. Breathing movement could be described by observing thoracoabdominal movement which is a combination of cooperated muscle contraction from rib cage and abdomen. Isovolume maneuver (IVM) is a typical respiratory pattern with different respiratory frequency and depth. To investigate the effect of respiratory pattern, it is important to observe the transition state between different respiratory modulations. In 2016, Huang proposed Holo-Hilbert spectral analysis (HHSA) for identifying the joint frequency and amplitude modulation. To obtain the transition period, HHSA is adopted to investigate time-frequency modulation (Time-FM) and time-amplitude modulation (Time-AM) during IVM. Simulation and experiment results indicate that time-AM spectrum and FM-AM spectrum can identify the time-variate transition period between spontaneous breathing (0.3 Hz) and paced breathing (0.1 Hz). This achievement could be helpful for detecting transition state of respiration, especially during exercises or suffering cardiopulmonary disease for long-term care.**

Keywords-Holo-Hilbert spectral analysis (HHSA); Empirical mode decomposition (EMD); Isovolume manuver (IVM)

I. INTRODUCTION

Breathing is an essential and important process for gas exchange. The mechanism of human breathing is the pressure gradient of airflow to induce inhaling or exhaling with a series of cooperation between ribs and muscles [1][2]. People can modulate instantaneous respiratory frequency (fast or slow) and depth (deep or shallow) consciously or unconsciously. There are two common breathing types, thoracic breathing (TB) and abdominal breathing (AB). TB is implemented by the contraction of intracostal muscles to raise rib cage and to increase lateral and anteroposterior dimension of thoracic cavity. Most of gas exchange happens in upper lungs. AB is commonly used during sleep and mainly implemented by the contraction of abdominal muscles. It would drive the deeper descending of the diaphragm and inhale air into the entire lungs. The lung volume during AB is usually larger than TB and performs slowly rate and more effective on gas exchange. In addition, different breathing types, especially AB, could stimulate parasympathetic nervous system and lead human to feel relaxed [3][4].

In physiological study, the dynamic function of breathing can be described by observing thoracoabdominal movement (TAM) consisting of related muscular contractions and relaxations between rib cage (RC) and abdomen (ABD) [5]. One of non-invasive methods to measure breathing pattern is respiratory inductive plethysmograph (RIP). In general, RIP belts are placed on the RC and ABD to measure the TAM and quantify the breathing features including respiratory rate, tilde volume, energy of breathing, and phase shift of TAM [6]. Literature has reported that patterns of RC and ABD movements can be used to identify breathing types (TB and AB) [7][8]. However, breathing pattern is dynamical changes, it is important to observe the transition state between different respiratory modulations.

In signal processing, noise reduction with band-pass filter has time delay problem. Empirical mode decomposition (EMD) [9] is used to decompose biosignals into oscillatory modes with different frequencies without introducing any phase delay [10]. Each component decomposed by EMD is regarded as a meaningful mode called intrinsic mode function (IMF). The instantaneous frequency responses of the IMFs derived through direct quadrature and showed in Hilbert spectrum [11]. In EMD process, mode mixing has seemed to be a problem, which is defined as a pattern of signals whose activities reside within the same frequency of different IMFs. In a new method, named Holo-Hilbert spectral analysis (HHSA) and proposed by Huang *et. al.* in 2016, there exists a physical phenomenon in the mode mixing [12]. It extends EMD method and uses nested EMD with Hilbert-Huang Transform (HHT) to identify the modulations in nonlinear systems [12]. In addition to frequency variation, data containing amplitude variation would influence EMD process. HHSA brings out a high-dimensional spectrum called Holo-Hilbert spectrum to see the variation in frequency and amplitude.

The iso-volume maneuver (IVM), which is widely used for calibrating respiratory sensors and thoracoabdominal movement evaluation [13][14], is a typical respiratory pattern with spontaneous breathing and paced breathing. To investigate the effect of respiratory pattern, it is important to observe the transition state between different respiratory modulations. HHSA has been proposed recently to identify the frequency and amplitude modulations in nonlinear systems. This study used HHSA to investigate time-frequency modulation (Time-FM) and time-amplitude modulation (Time-AM) during IVM.

II. METHOD AND METERIAL

A. *Holo-Hilbert spectral analysis method*

HHSA extends current spectral analysis such as Fourier analysis and Hilbert spectral analysis, which are based on additive expansion to show the variation on frequency. It gets the lower frequency envelop as amplitude fluctuation from the higher carrier frequency oscillations and expands original time-frequency spectrum into a higher-dimensional representation of the FM-AM spectrum, with the FM representing the fast-changing carrier intra-mode frequency variations and the AM representing the slow-changing envelope inter-mode frequency variations [12]. A best way to present the advantage in HHSA as shown in (1). A multiplicative data would be regarded as the addition from two terms by using an additively based method. On the other hand, it could be considered as a cosB term with the amplitude modulation by cosA term. Therefore, we could obtain the amplitude variation through HHSA if it exists amplitude modulation.

$$x(t) = \cos A \cdot \cos B = (1/2)\,[\cos(A+B)+\cos(A-B)] \quad (1)$$

Before introducing HHSA method, realized the EMD algorithm ahead. The steps of EMD are shown in Fig. 1(a). First, we can find the local maximum and local minimum to produce upper envelope and lower envelope by a cubic spline and calculate the mean of these two envelopes. Then, subtract the mean envelope from the oringinal data. The above detrending operation is called sifting process, which is based on energy-associated extraction in each timescale [9]. After sifting process, we need to determine whether the result is an intrinsic mode function (IMF). An IMF would satisfy the following two requirements: (1) the number of local extrema must be equal or differ at most by one to the number of zero-crossing; (2) at any point, the value of the mean envelope must approximately equal to zero. If the

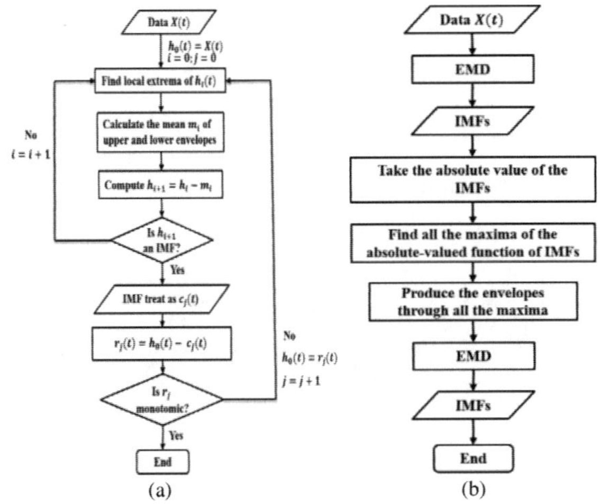

Figure 1. Illustration to the procedure of (a) EMD algorithm and (b) HHSA.

result is an IMF, output the IMF and deduct it from the original data to get the residue. If not, replace the original data by the result and perform sifting process again. Last, determine whether the residue is monotonic. If it is not a monotonic function, use the residue as original data and execute above steps. Else, EMD procedure is finished and gets lots of IMFs decomposed by EMD method.

The steps of HHSA method are shown in Fig. 1(b). Data would be decomposed by EMD method more than one time. First, original data execute EMD method and obtain many IMFs, treating as IMF_i that means i-th IMF. We take the absolute value on each IMF_i and identify all the maxima on it to produce envelope by a natural spline which be viewed as the time fluctuation of the amplitude function. Then, perform EMD method on these envelopes and produce another lots of IMFs, treating as IMF_{ij} that means j-th IMF in the envelop of IMF_i. The step, decomposing envelopes by EMD method, would constantly execute to delineate the amplitude function. With each additional layer of decomposition on the envelopes, need additional dimensions to accommodate the amplitude variations. This study presents two layers EMD method in HHSA and gets Holo-Hilbert spectrum with four – dimensional data including time, FM frequency, AM frequency, and energy density. Besides the FM-AM spectrum, defined in [12], we show the Time-FM spectrum and Time-AM spectrum to investigate FM frequency variation and AM frequency variation on breathing pattern in timescale.

B. *Simulation and IVM data collection*

This study simulated breathing pattern by using pure sinusolid wave and showed two different breathing types alternating appeared as Fig. 2. One breathing rate is 20 times per minute (0.3 Hz). The other one is 6 times per minute (0.1 Hz) and has larger amplitude than the amplitude of 0.3 Hz breathing rate. The breathing pattern executes five times in simulation.

Procedure of IVM includes alternate spontaneous breathing (near 0.3 Hz) and paced breathing (0.1 Hz) in

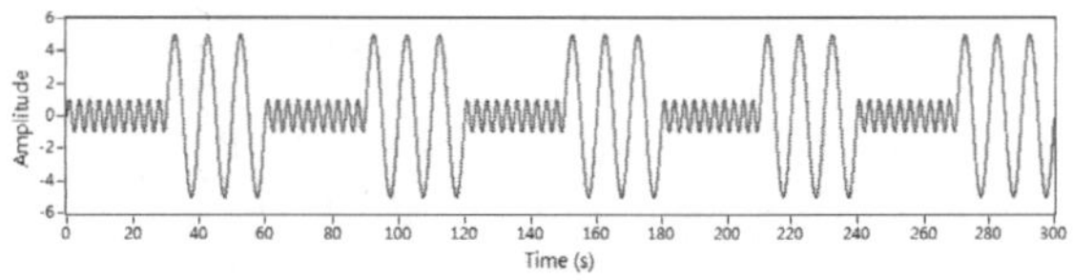

Figure 2. Simulation of spontaneous breathing (0.3 Hz) and paced breathing (0.1 Hz) with sinusoidal pattern.

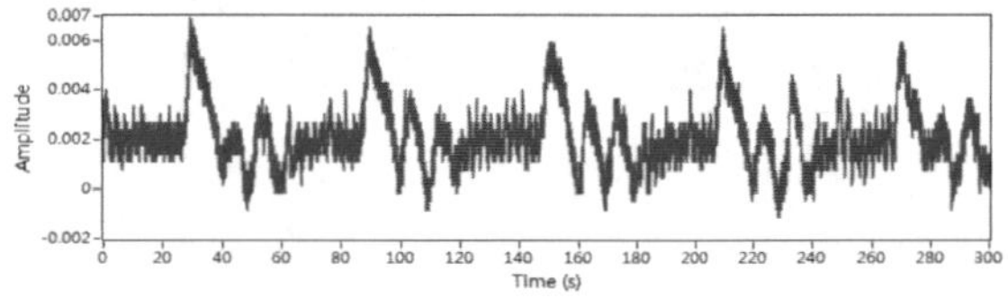

(a)

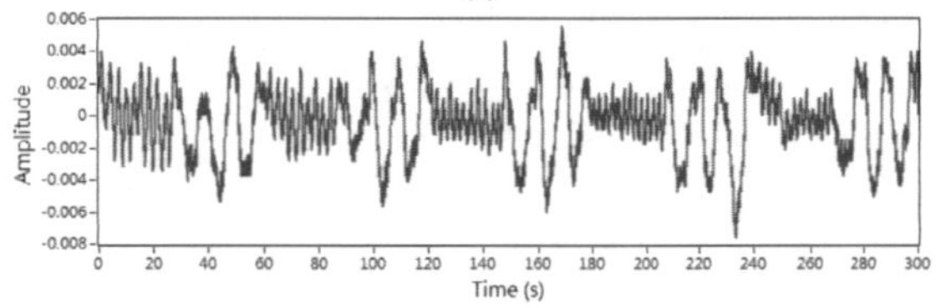

(b)

Figure 3. RIP signals from (a) RC movement and (b) ABD movement during IVM. The dense part is for spontaneous breathing and the loosely part is for paced breathing.

iso-volume status. Subject (female, 24 years old, 165 cm, 50kg) was instructed to perform IVM as following four steps for five times. (1) The subject breathed freely for 24 seconds and prepared to start the IVM. (2) The subject was asked to inhale deeply and slowly to hold the breath for 6 seconds. The lung volume kept equivalent during the IVM. (3) The subject was instructed to move abdominal wall inward for 5 seconds and moved outward for 5 seconds. The abdominal wall moved inward and outward for three times. (4) The subject was asked to exhale and take a break for 24 seconds. During IVM, shift air between RC and ABD as much as possible with the holding breath.

Fig. 3 illustrates the data, which were be acquired by two RIP sensors (RIPmate Adult Thorax Alice 5 Inductance Kit, Ambu Inc., Denmark) during IVM procedure. The RIP sensor was worn below the axilla to record RC movement as shown in Fig. 3(a), and the other RIP sensor was placed on the navel to record ABD movement as shown in Fig. 3(b). Both RIP signals were acquired by a data acquisition hub (NI SCB-68, National Instruments, USA) and a data acquisition card (NI USB 6255, National Instruments, USA) at a sampling rate of 50 Hz and were subsequently transferred to a computer (Acer Veriton M2610). The program controlling the instructions and data acquisition was developed using LabVIEW platform (LabVIEW 2012, National Instruments, USA).

III. RESULTS

Fig. 4(a) illustrates Hilbert spectrum with EMD method and instantaneous frequency in simulation data. Fig. 4(b), 4(c), and 4(d) illustrate HHSA method including FM-AM spectrum, Time-FM spectrum, and Time-AM spectrum respectively. In Hilbert spectrum, we observe the alternate frequency, 0.33Hz and 0.1 Hz, in timescale which is consistent with the condition we gave. Because the magnitude is calculated by energy density, larger amplitude indicates as high magnitude and red color in this case. In FM-AM spectrum, it presents high energy (high magnitude) when FM frequency is equal to 0.1Hz and 0.33 Hz and obtain high energy when AM frequency is within the range of 0.015 to 0.02 Hz. In Time-FM spectrum, the distribution of frequency is similar to Hilbert spectrum but different in color. In Time-AM spectrum, we find the high energy at about 0.015 Hz in AM frequency and appear high frequency (0.09 Hz) near transition timing.

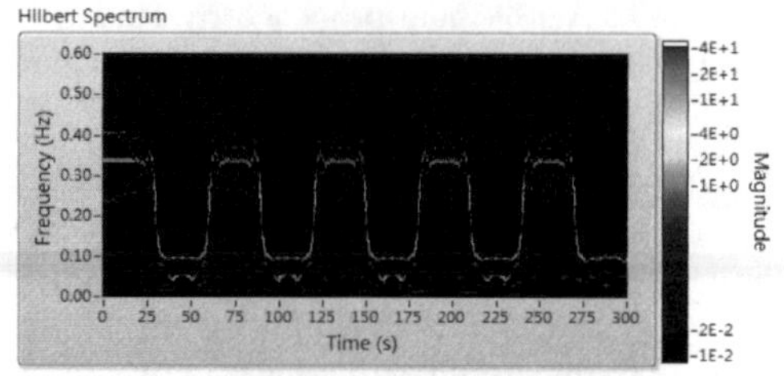

(a)

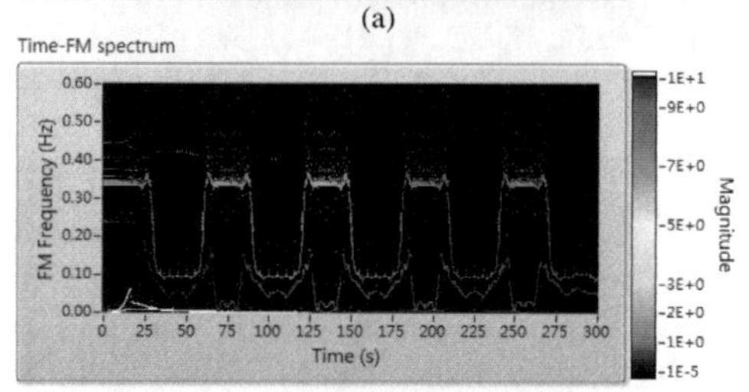

(b)

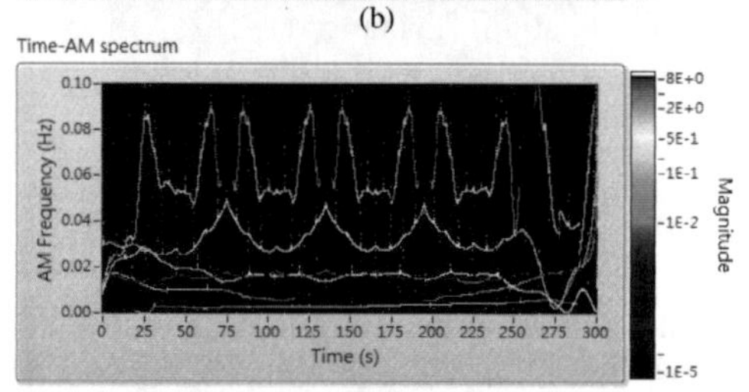

(c)

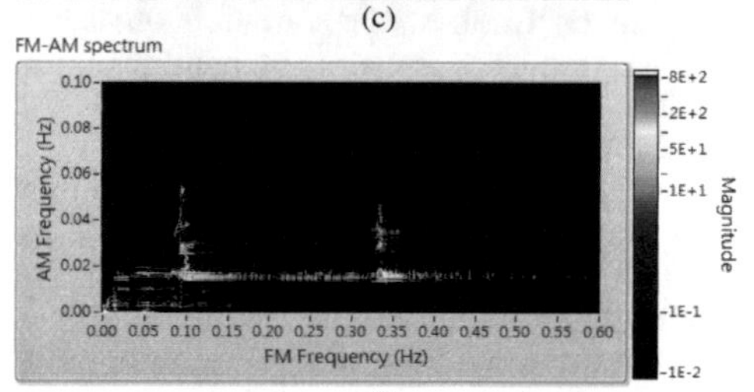

(d)

Figure 4. The spectra for the simulating data consist of (a) Hilbert spectrum, (b) Time-FM spectrum, (c) Time-AM spectrum, and (d) FM-AM spectrum.

Fig. 5 and Fig. 6 indicate the results of RIP signals from RC and ABD movement. We see RC movement first. In Hilbert spectrum, it appears the floating frequency with time. The frequency in spontaneous breathing is about 0.2 to 0.4

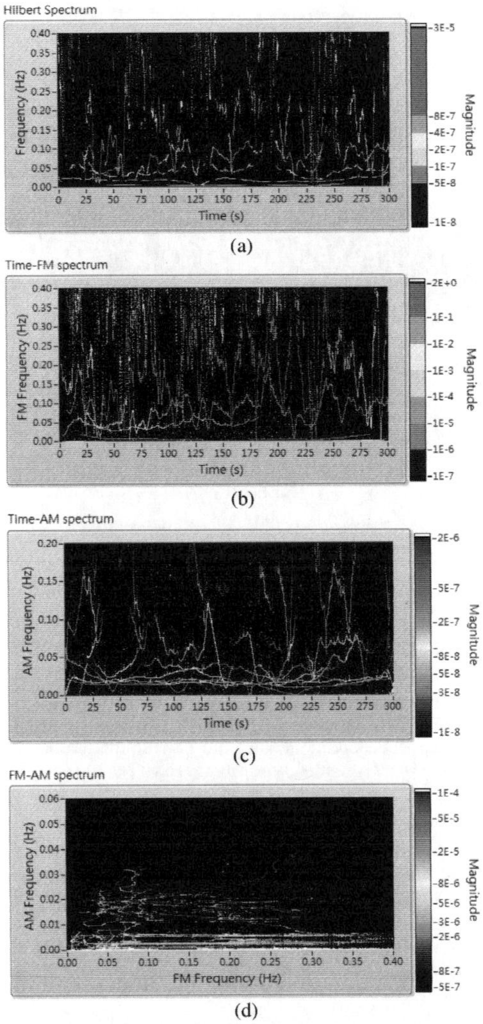

Figure 5. The spectra for the RIP signal with RC movement consist of (a) Hilbert spectrum, (b) Time-FM spectrum, (c) Time-AM spectrum, and (d) FM-AM spectrum.

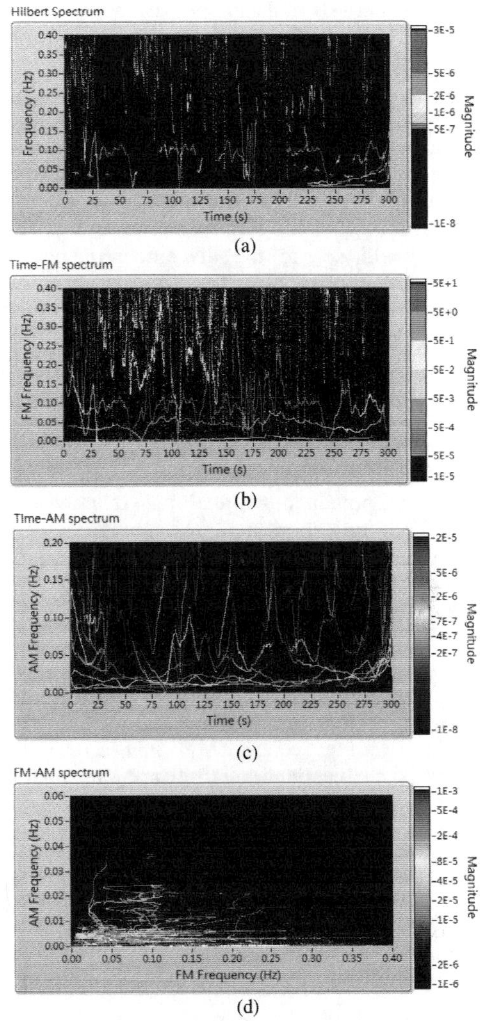

Figure 6. The spectra for the RIP signal with ABD movement consist of (a) Hilbert spectrum, (b) Time-FM spectrum, (c) Time-AM spectrum, and (d) FM-AM spectrum.

Hz with yellow and green color and the frequency in paced breathing range from 0.05 to 0.15 Hz with red color (higher energy). In FM-AM spectrum, there are two faint parts in FM frequency at 0.05 to 0.1 Hz and 0.15 to 0.4Hz respectively and appears higher energy near 0.02 Hz in AM frequency. In Time-FM spectrum, more complex than simulating data, we could recognize three distinct ranges. One of them is in the range of 0.05 to 0.1 Hz with green and blue color that is viewed as the frequency of paced breathing. Another is ranging from 0.2 to 0.4 Hz with yellow and green colors that refers to the frequency of spontaneous breathing. The other one is most obvious that joined both of above and crossed the range from 0.05 to 0.4 Hz. In Time-AM spectrum, there is a clear orange line in 0.02Hz but others information is irregular.

The spectra in ABD data compare to RC data, there are some difference in FM-AM spectrum and Time-FM spectrum. In FM-AM spectrum, the energy in lower frequency is larger and more centralized than RC in FM frequency. In Time-FM spectrum, the red region shows lower frequency than RC.

IV. DISCUSSION

We performed the simulated breathing data on four-type spectra and found that the similar distribution of frequency in Hilbert spectrum and Time-FM spectrum expect of 0.1 Hz and 0.3 Hz components allocated different colors, implying different energy densities. Compared to Hilbert spectrum, the pure frequency variation is clearly displayed in Time-FM spectrum. This performance could be the effect of 2nd EMD process, displaying it in another Time-AM spectrum. In addition, we find two bright points near 0.02Hz in AM frequency in FM-AM spectrum. AM is the slow changing envelope frequency variation that presents cyclic characteristics in the envelopes. In simulation, breathing pattern executed five times could be regarded as transition period with five periods in the envelopes and its frequency is 0.016 Hz. It is close to the higher energy in AM frequency indicated in FM-AM spectrum and Time-AM spectrum.

Comparing real breathing pattern with simulating breathing pattern and RC movement with ABD movement, we find some results. (1) There are two cases under the highly varied frequency of spontaneous breathing. In simulating case, we could clearly see the frequency in Hilbert spectrum and Time-FM spectrum. But in real case, the energy would spread to adjacent frequency, so the frequency is ambiguous. (2) In Time-AM spectrum, we find the higher frequency timing close to the transition timing in simulating case, but could not find in real case fellow the same rule. It may owing to the energy of higher frequency pretty lower than main energy that is difficult to indicate clearly in dynamical respiration. (3) The main distribution of frequency on ABD movement is lower than RC movement in Time-FM spectrum that present higher energy on paced breathing than spontaneous breathing. It may cause by performing TB during spontaneous breathing but we need to more experiment data to prove. (4) There are some unreasonable distribution of energy in Time-FM spectrum. As the Fig. 5(b) shows that the main distribution of energy range from 0.05 to 0.4 Hz, it is composed by two kind of frequency of breathing pattern. After being decomposed by EMD method, it should be divided into two different frequencies.

The IMFs decomposed by EMD on RC data shown as Fig. 7(a). We find the mode mixing problem in IMF_6. We repeat the same process after removing IMF_6 from RC data. In Time-FM spectrum shown as Fig. 7(c), we can find 0.1 Hz frequency of paced breathing obviously and 0.2 to 0.4 Hz in spontaneous breathing. Followed by above steps on ABD data, we can obtain clear frequency in spontaneous breathing and paced breathing.

V. CONCLUSION

In this study, we have used HHSA method on simulation of breathing pattern and respiratory movement data with IVM procedure. The frequency of amplitude fluctuation can be found in both Time-AM spectrum and FM-AM spectrum. Therefore, the transition period between spontaneous breathing (0.3 Hz) and paced breathing (0.1 Hz) can be found in Holo-Hilbert spectrum. Based on the HHSA method, we got the FM frequency and AM frequency in timescale to investigate the transition from spontaneous breathing to paced breathing. Related results are helpful for exploring the transition state on human breathing and dynamical respiratory pattern especially when people doing exercises or suffering cardiopulmonary disease for long-term care.

ACKNOWLEDGMENT

This work was fully supported by the Taiwan Ministry of Science and Technology under grant numbers MOST 105-2634-E-009-003 and MOST 105-2221-E-009-159. This work was also supported in part by the "Aim for the Top University Plan" of Biomedical Electronics Translational Research Center in National Chiao Tung University and Ministry of Education, Taiwan.

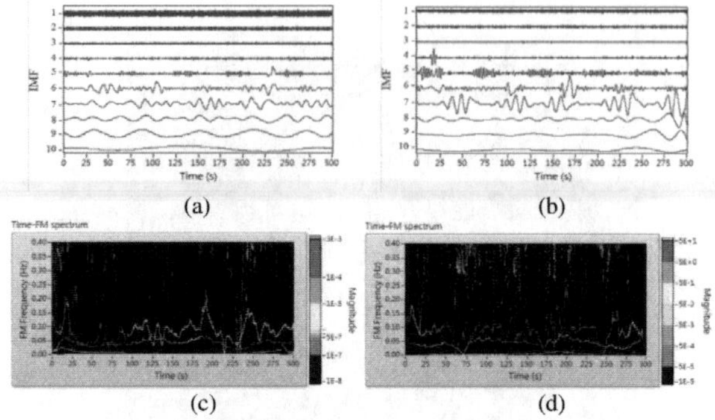

Figure 7. (a) The IMFs for the real data on RC. (b) The IMFs for the real data on ABD. (c) Time-FM spectrum of RC data removing IMF_6. (d) Time-FM spectrum of ABD data removing IMF_6.

REFERENCES

[1] B. W. John, *Respiratory Physiology: the essentials*. Alphen aan den Rijn: Wolters Kluwer, 2012.

[2] W. F. Ganong, *Review of Medical Physiology*, 22 ed.: Mcgraw-hill, 2005.

[3] A. Conrad, *et al.*, "Psychophysiological effects of breathing instructions for stress management," *Applied Psychophysiology and Biofeedback*, vol. 32, pp. 89-98, 2007.

[4] R. Kaushik, R. M. Kaushik, S. K. Mahajan, and V. Rajesh, "Biofeedback assisted diaphragmatic breathing and systematic relaxation versus propranolol in long term prophylaxis of migraine," *Complement. Ther. Med.*, vol. 13, pp. 165-174, 2005.

[5] K. Konno and J. Mead, "Measurement of the separate volume changes of rib cage and abdomen during breathing," *Journal of Applied Physiology*, vol. 22, pp. 407-422, 1967.

[6] J. Hammer and C. J. L. Newth, "Assessment of thoraco-abdominal asynchrony," *Paediatr Respir Rev*, vol. 10, pp. 75-80, 2009.

[7] J.-L. Chen, Y.-C. Chen, and T.-C. Hsiao, "Recognizing thoracic breathing by ensemble empirical mode decomposition," *in Information, Communications and Signal Processing (ICICS) 2013 9th International Conference on*, pp. 1-5, 2013.

[8] Y.-C. Chen and T.-C. Hsiao, "Instantaneous phase difference analysis between thoracic and abdominal movement signals based on complementary ensemble empirical mode decomposition," *BioMedical Engineering OnLine*, vol. 15, pp. 1-21, 2016.

[9] N. E. Huang, *et al.*, "The empirical mode decomposition and the Hilbert spectrum for nonlinear and non-stationary time series analysis," *Proc. R. Soc. Lond. A*, vol. 454, pp. 903-995, 1998.

[10] C.-C. Chang, S.-C. Kao, T.-C. Hsiao, and H.-Y. Hsu, "Assessment of autonomic nervous system by using empirical mode decomposition-based reflection wave analysis during non-stationary conditions," *Physiological measurement*, vol. 35, pp. 1873-1883, 2014.

[11] N. E. Huang, *et al.*, "On instantaneous frequency," *Adv. Adapt. Data Anal.*, vol. 1, pp. 177-229, 2009.

[12] N. E. Huang, *et al.*, "On Holo-Hilbert spectral analysis: a full informational spectral representation for nonlinear and non-stationary data," *Proc. R. Soc. Lond. A*, vol. 374, pp. 1-21, 2016.

[13] R. C. C. Barbosa, C. R. F. d. Carvalho, and H. T. Moriya, "Respiratory inductive plethysmography: a comparative study between isovolume maneuver calibration and qualitative diagnostic calibration in healthy volunteers assessed in different positions," *Jornal Brasileiro de Pneumologia*, vol. 38, pp. 194-201, 2012.

[14] R. Gilbert, J. Auchincloss, and D. Peppi, "Relationship of rib cage and abdomen motion to diaphragm function during quiet breathing," *Chest Journal*, vol. 80, pp. 607-612, 1981.

Engaged Scholarship as Research Method: a Best Practice for evidence based IT-innovations for People with Severe Dementia

Elles Gyaltsen-Lohuis, Marike Hettinga
Research group IT Innovations in Health Care
Windesheim University of Applied Sciences
Zwolle, The Netherlands
Email: {EGE.Lohuis} {M.Hettinga}@ windesheim.nl

Abstract— **Providing evidence for the added value of IT-innovations in health care for people with dementia is most challenging. Questions on relevant outcome criteria, methods of measurement, qualitative methods and quantitative, and correct stakeholder management are some of the issues coming to the fore. This short paper shares the best practices and challenges met on two short-term studies on IT-innovations for people with severe dementia. The paper also includes the topics and questions to discuss in the workshop EviDem: Evidence based IT-innovations for People with Dementia, during the eTelemed, the Ninth International Conference on eHealth, Telemedicine, and Social Medicine, 2017.**

Keywords: engaged scholarship; IT-innovations in health care; dementia; research method; evidence.

I. Introduction

In 2015 and 2016 I conducted two studies on IT innovation in intramural care for people with severe dementia. The first study was on the effects of a remote sensor system as regards to 1. the quality of life and care of the client and 2. the efficiency of care by the care providers. The second study was on the effects of a social media application as regards to 1. the quality of life of the client and 2. the relationship between client and care provider. Both studies were small-scale, involving 12 to 24 clients with severe dementia and lasted six months. This short paper focusses on the methods used and the challenges met.

II. Engaged Scholarship and Ethnography as Research Methodology

The study on IT innovation for people with dementia in an intramural setting focuses on a series of complex issues in a social setting. The IT innovation is supposed to generate different results for various stakeholders in the arena of care, and all stakeholders have different interests. Clients with dementia are mostly hoping the IT innovation will improve their quality of life and care. Formal care providers often want to be able to give better care with the IT innovation. Boards of management of care providers direct their attention to efficiency in care while IT innovators and also financers of the IT innovation have commercial stakes when it comes to innovation in care practices.

At the same time the study on IT innovations is also supposed to contribute to the body of scientific knowledge on design and implementation of IT innovation in care. So the research method used has to serve multiple stakeholders in both the practice and the science of care. To increase the capabilities to study complex problems and to create the kind of knowledge that advances both practice and science in care, I propose the method of engaged scholarship by Van de Ven [1]. Engaged scholarship is a participative form of research for obtaining the views of key stakeholders to understand a complex problem. By exploiting differences between the viewpoints of the stakeholders, I found that engaged scholarship produces knowledge that is more penetrating and insightful than when researchers work alone.

Depending on the research purpose Van de Ven proposes 2 possible research perspectives: the 'detached outsider' and the 'attached insider' [1]. The purpose of my research was two-fold, namely to gain knowledge about the effects and the adaption of the IT innovation interventions together with the stakeholders and to gain knowledge about the complexity of intramural setting in which the IT innovation is launched. Therefore I chose the 'attached insider' view.

The first aim of studying the effects and adaption of the IT innovation demanded an action/intervention research. Developed by Kurt Lewin in 1944 action research is a reflective process of problem solving whereby researchers and stakeholders work together to improve the community's strategy, practices and knowledge of the settings in which they practice. The basis of this process is the participation of stakeholders in problem solving, using systematic methods of data collection, feedback, reflection and action [1]. By taking the perspective of the 'attached insider', I worked together with the key stakeholders to design and execute the research to ensure that all stakeholder's interests were served the best possible way.

The second aim, to gain insight in the complexity of the situation and to contribute to the body of scientific knowledge, required a collaborative basic research, another form of engaged scholarship [1]. This form of research entails sharing of power and joint activities among the researcher and stakeholders in order to co-produce knowledge about a complex problem or phenomenon. Taking again the perspective of the 'attached insider', I shared the daily life with the clients and care providers in the intramural setting for a period of time, discussing our experiences and our viewpoints, their perceived obstacles and analyzing our findings together. I gathered the experiences and knowledge of the diverse stakeholders involved through interviews, conversation, participation and observation. To get a close and intimate familiarity with the stakeholders and their practices the key in the research strategy used here was participant observation, observing while playing a role within the group observed.

III. CHALLENGES MET

As Van de Ven notes, practicing engaged scholarship raises a number of challenges, two of which I also encountered while doing research in the intramural setting: divergent viewpoints and building relationships with stakeholders. Firstly, engagement generates divergent viewpoints that can be juxtaposing. In my research I incorporated the stakeholders' viewpoints and I gathered data from all the stakeholders involved: the clients, their care providers – formal and informal -, the board of management, the IT innovation developers and the financers of the IT innovation. The viewpoints generated showed different stances – not only on the IT innovation itself as intervention but also on the results this IT innovation generated by the client with dementia.

I found that reconciling these divergent viewpoints is not the point in this research. The point is to study and understand when and why these viewpoints are different. I tried to explain these differences by seeing the interconnectedness and webs of entanglement between the different dimensions of the processes in the daily lives of the clients and their care providers – formal and informal. I came to a holistic understanding from engaging all the stakeholders and analyzing their constantly evolving relationships.

The second challenge was negotiating the research relationship by establishing and building relationships with stakeholders. My research was undertaken with all stakeholders but especially I needed to build a good relationship with the clients and their formal care providers with whom I would be working. I experienced that building this relationship takes time, mutual understanding and a subordinate position of the researcher. The clients and care providers in my research were never involved in any sort of research before. When I suggested this research to them I made sure that they understood what it entailed from both sides, coming to work with them, talking and sharing, explaining, reflecting and exposing personal details.

The majority of the clients involved could not communicate and understand the implications of being involved in research. Their interests were represented by the care providers – formal and informal. The board, informal and the formal care providers had outlined their priorities as regards the clients: the wellbeing of the client at all stages of the research. This meant no intrusion, no upheaval, no insecurities in the clients' environment. As soon as it became clear that the client had a negative reaction or was obstructed in his or her daily life, the research and IT intervention would be stopped. Keeping the daily routine as regular as possible was key to the wellbeing of the clients. We made the agreement that the clients' wellbeing would always have precedent over any other issue that could in the study. All decisions made during the research were always guided by these principles. This way we made sure that all parties were contributors in working towards the development of the IT innovation and partners in the study.

Taking the time to build the relationship, mutual understanding and a complete subordinate position from the researcher to the wellbeing of the client made all stakeholders feel secure and safe, contributing as much as what was within their reach and therefore ensuring the best results within the study. It also meant however that the length of the study and the schedule set was determined by the clients from day to day. For me as a researcher this imposed a definite time challenge in the research frame that was dictated by strict delivery dates and deadlines.

IV. TOPICS AND QUESTIONS TO DISCUSS

The other challenges that I experienced in my studies on IT innovations and people with severe dementia are: 1. how to include the persons with severe dementia in the IT development and 2. how to take the constantly changing needs and abilities for persons with severe dementia into account while ensuring reliable research data.

It is proven that including the persons with dementia in the development of IT innovations ensures a better product/service and a higher adaptation of the innovation [2]. However, people with severe dementia are often not able to participate in research due to the fact that dementia in a severe stage restricts their interaction with their environment. To ensure the best results, what is the best practice to include the persons with severe dementia in the IT development?

During the research I experienced that the ability to interact with people with severe dementia can vary from hour to hour. One moment the person with severe dementia is able to interact and use the IT innovation provided, the next moment he or she is not able to due to the effects of the illness or the effects of the medication administered. This is an incredible challenge to both the client and the researcher. What is the best practice to take the constantly changing needs and abilities for persons with severe dementia into account while ensuring reliable research data?

REFERENCES

[1] A.H. Van der Ven, Engaged Scholarship: A Guide for Organisational and Social Research. Oxford:: University Press, 2007.

[2] M. Span, Developing an Interactive Web Tool to Facilitate Shared-Decision Making in Dementia Care Networks: A Participatory Journey. [Online]. Available from: http://hdl.handle.net/1871/54018

Simulation-Based Learning in Undergraduate Nursing Education in Japan

A review of the literature

Kazuma Takahashi,
Natsuko Miura,
Department of Basic Nursing Science,
Faculty of Nursing, Graduate School of Nursing, Iwate Prefectural University,
Takizawa, Japan.
e-mail: t_kazuma@iwate-pu.ac.jp, natsuko@iwate-pu.ac.jp

Abstract—**The use of simulation in health professional education has increased rapidly in Japan during the past decade. However, to date, only a limited number of studies reporting this situation has been published. We, herein, endeavored to ascertain how simulation has been adopted in nursing education, reviewing Japanese articles in this field. Searching a database, Nippon Igaku Chuo Zasshi, identified 86 peer-reviewed research studies published in Japanese between 1986 and 2016. Since 1999, around 5 articles per year have consistently been published. Simulation was applied predominantly in basic and medical-surgical nursing areas, to learn physical assessment, fundamental skills, injection, and perioperative care. Forty-one percent of the simulation-based learning was conducted using 3-D-models to provide infection and cardiopulmonary resuscitation education. Intermediate and high fidelity patient simulators were applied to teach physical assessment, fundamental skills, and perioperative care, only in rare cases. In Japan, much higher importance should be placed on simulation-based undergraduate nursing education, and the socioeconomic environment for introducing more sophisticated simulators must be improved.**

Key words- nursing education; nursing; simulation; simulators; health care professionals.

I. Introduction

The primary aim of simulation is to improve patient safety and to help the student nurse achieve competence, linking their theoretical knowledge with clinical practice [1]. The use of simulation in health professional education has increased rapidly over the past 2 decades [2]. The current range of simulation in this field consists of written simulations, three-dimensional models, screen-based simulators, standardized patients (SP), intermediate fidelity patient simulators, and interactive patient simulators (Table 1)[3].

In the current health education system, the demand for clinical placements is increased, while there is a global social need for higher quality healthcare services with shorter hospital stays [4]. In addition, in Japan, clinical practicum hours have been decreasing remarkably in the recent nursing education curriculum, as students spend more time in theoretical classroom learning. Thus, nursing students have less chance to be involved in patient care and reduced opportunities to deal with practice situations [4]. This may lead to decreased clinical competencies in nursing school graduates [5]. Hence, there has been a need to reproduce the experience by some other means. Indeed, the Japanese Ministry of Health, Labour and Welfare encouraged trainees to complement skills that could not be experienced during a practicum using simulation-based learning.

Reflecting a global trend of accelerating use of simulation in nursing education, the simulation-based education in our country seems to be spreading recently. However, a limited number of studies going through this situation has been published in Japanese, and there have been none in English. In this study, we endeavored to ascertain how simulation has been adopted in nursing education in Japan, reviewing Japanese articles in this field.

II. Methods

A. Literature Search Strategies

The research question that informed the search was 'How has simulation been used in nursing education in Japan?' Boolean operators of AND & OR were used in the search. A Japanese electronic database, Nippon Igaku Chuo Zasshi (Japana Centra Revuo Medicina), was searched for peer-reviewed research studies in the Japanese language published between the years of 1986 and December 2016, focusing on simulation and/or simulator use in nursing education. The search terms used to discover the literature were: Simulation; Simulators; Nursing; Education; Skills.

B. Inclusion and Exclusion Criteria

Studies were selected if they described an impact of simulation-based learning in nursing undergraduate education. Excluded studies were literature reviews, those reviewing learning theory, qualitative reviews of student or faculty experience, and those targeting simulation education for qualified staff.

C. Literature review and analysis

The searches identified a total of 86 articles on simulation-based undergraduate nursing education during the period. Articles were grouped based on the major and sub-specialties of nursing science, types of simulation, and subjects who were the targets of simulation training. The

TABLE I. PROPOSED TYPOLOGY OF SIMULATION METHODOLOGIES BY ALINIER [3]

Simulation Techniques	Examples
Written simulations	Pen and paper simulations, or "Patient Management Problems", and latent images
3-D models	Basic mannequins, low fidelity simulation models, or part-task simulators
Screen-based simulators	Computer simulation, simulation software, videos, DVDs, or Virtual Reality and surgical simulators
Standardized patients	Real or simulated patients, role playing
Intermediate fidelity patient simulators	Computer controlled, programmable full body size patient simulation not fully interactive
Interactive patient simulators or computer controlled model driven patient simulators	High fidelity simulation platforms

major specialties included basic (fundamental nursing skills), communities/public, geriatric, medical-surgical, midwifery, pediatric, and psychiatric and mental health nursing, and the subspecialties of cardiac, critical care, disaster, emergency, home, informatics, management, oncology, orthopedic, perioperative, and wound, ostomy and continence (WOCN) nursing, in addition to major categories. Studies dealing with general subjects of major specialties were sub-classified into the category designated as being as the same as the major one (Table 2).

III. RESULTS

A. Annual Change in the Number of Articles on Simulation-based Undergraduate Nursing Education in Japan

The first report on simulation-based learning for nursing students appeared in 1987. Since 1999, the articles on simulation-based education for undergraduate nursing students have been consistently published, around 5 articles per year on average (Fig. 1).

B. Spcialities in Simulation-based Nurse Education

Simulation was applied predominantly to basic and medical-surgical nursing areas; the proportion of the basic nursing was 57%, and that of medical-surgical nursing 22%

TABLE II. NUMBER OF ARTICLES CLASSIFIED BY MAJOR AND SUB- SPECIALITIES

		Major spcialities							Total
		Basic	*Communities/ public*	*Geriatric*	*Medical-surgical*	*Midwifery*	*Pediatric*	*Psychiatric and mental health*	
Subspecialities	Basic	29	0	0	2	0	2	0	33
	Cardiac	1	0	0	0	0	0	0	1
	Critical care	0	0	0	2	0	0	0	2
	Disaster	2	0	0	0	0	0	0	2
	Emergency	6	0	0	1	1	0	0	8
	Geriatric	0	1	0	0	0	0	0	1
	Home	0	1	0	0	0	0	0	1
	Informatics	0	2	0	0	0	0	0	2
	Management	4	0	1	0	0	2	0	7
	Medical-surgical	2	0	0	1	0	0	0	3
	Midwifery	0	0	0	1	3	0	0	4
	Oncology	1	0	0	1	0	0	0	2
	Orthopedic	0	0	1	0	0	0	0	1
	Pediatric	2	0	0	0	0	0	0	2
	Perioperative	2	0	0	10	0	0	0	12
	Psychiatric and mental health	0	0	0	0	0	0	3	3
	WOCN	0	0	1	1	0	0	0	2
Total		49	4	3	19	4	4	3	86

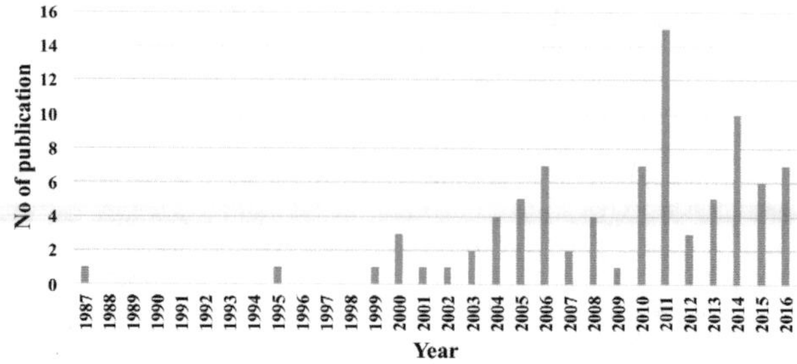

Figure 1. Annual Change in the Number of Articles Reporting Simulation-Based Undergraduate Nursing Education in Japan

(Fig. 2). Among the subspecialties, simulation was primarily used to learn basic nursing science (Fig. 3).

C. Training Subjects Intended in Simulation-based Nursing Education

Simulation was mainly implemented to educate trainees in physical assessment, fundamental skills, injection, and perioperative care (Fig. 4). The subject "injection" mostly included venipuncture.

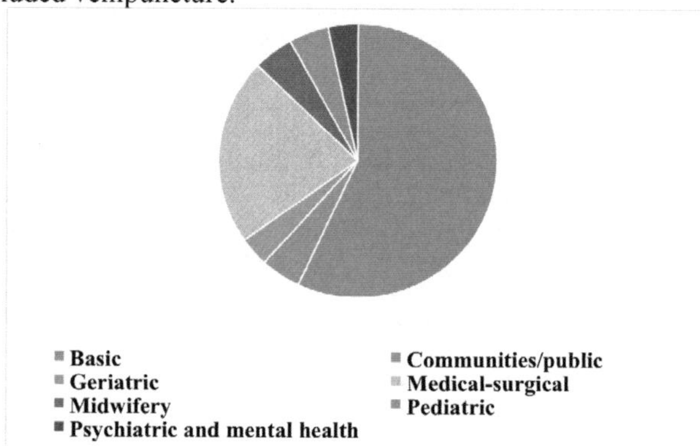

Figure 2. Major Specialties in Simulation-based Nurse Learning

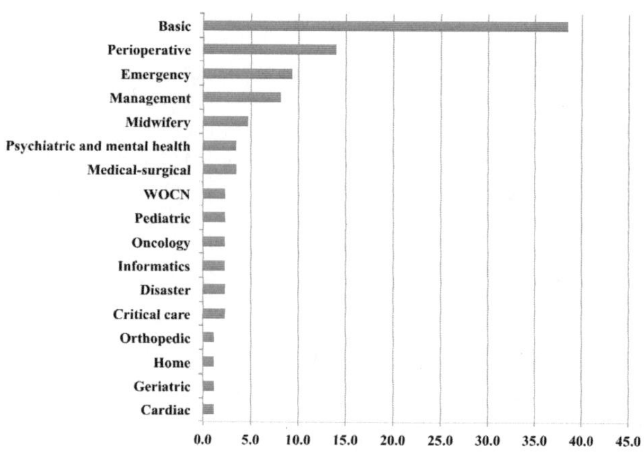

Figure 3. Subspecialties in Simulation-based Nurse Education

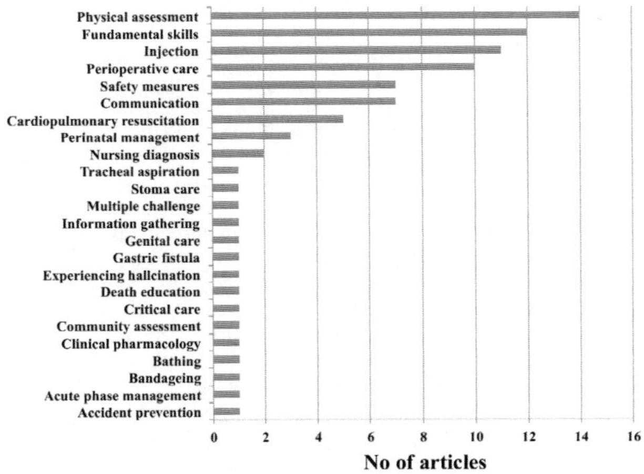

Figure 4. Training Subjects in Simulation-based Nursing Education

In the subspeciality "basic nursing", which comprised the main part of the simulation-based education (Fig. 3), the major training subjects consisted of injection, fundamental skills, physical assessment, and communication (21, 19, 12, and 9%, respectively).

D. Types of Simulation in Undergraduate Nursing Education

Forty-one percent of the simulation-based learning was conducted using 3-D-models (Fig. 5), to teach infection management and cardiopulmonary resuscitation (CPR) (Fig. 6). SP were used in communication and fundamental skills education (Fig. 6). Intermediate and high fidelity patient simulators were applied to learn physical assessment, fundamental skills, and perioperative care education, only in rare cases (Fig. 6). Articles targeting education with these sophisticated simulators first emerged in 2003, and have been consistently produced since 2010 (Fig. 7).

IV. DISCUSSION

To our knowledge, this is a first literature review in English reporting the simulation-based learning for undergraduate nurses in Japan.

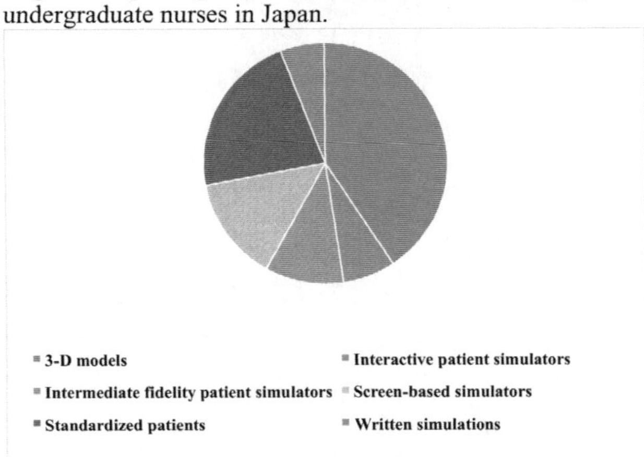

Figure 5. Types of Simulation

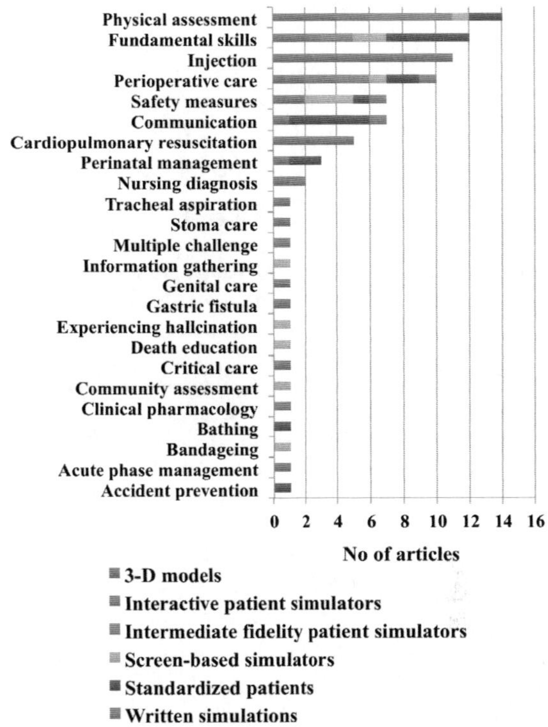

Figure 6. Distribution of Types of Simulation in Training Subjects

Since 2003, the number of articles in this field has started increasing, possibly highlighted and promoted by four governmental bulletins. The Japanese Ministry of Education released a report on the university/college education for practical nursing skills in 2002, and one on the attainment target for this education in 2004. In 2003, the Japanese Ministry of Health, Labour, and Welfare also issued a report on the desired skill education for undergraduate nursing students, and one on education in practical nursing skills for novice nurses. These noted that nursing students have fewer chances to be involved in patient care, and to deal with practice situations, because of a socioeconomic need for shorter hospital stays, and consideration of the human rights of patients, and medical safety measures. The bulletin in 2003 suggested applying role-playing to practical training within the campus, which may have pioneered simulation-based nursing education in Japan. Indeed, SD was one of the major issues in the articles published (Fig. 5), and these studies came out consistently (Fig. 7).

This literature review revealed that 3-D models, including basic mannequins, low fidelity simulation models, or task trainers were prevalently used in pre-licensure nursing education in Japan, when teaching physical assessment, general fundamental skills, injection, or CPR. This observation, although literature based, is in line with an actual survey by Kuroda et al [7], showing that 61.2 % and 69.3 % of nursing schools possessed task trainers and low fidelity simulators, respectively.

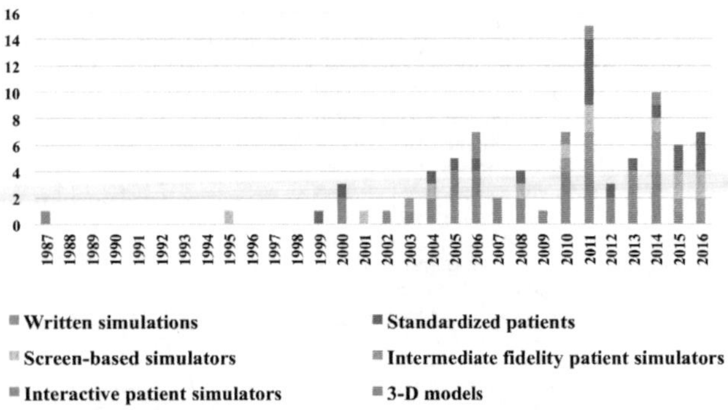

Figure 7. Annual Changes in Types of Simulation

A report reconsidering the teaching contents and methods in the undergraduate nursing education released by the Japanese Ministry of Health, Labour, and Welfare in 2012 advised that skills, which could not be experienced during the practicum training, should be complemented by simulation, and even referred to introducing more sophisticated simulators, which may have brought about emergence of the articles addressing intermediate and high fidelity simulators (Fig. 7). Multiple studies confirmed that high fidelity simulation benefited nursing students in terms of knowledge acquisition and critical thinking, and created a learning environment that contributed to greater knowledge, skills, safety, and student confidence [7]. These devices, however, have not been sufficiently distributed in our country, to date; the ownership ratio of intermediate fidelity simulators for physical assessment, that for critical care, and high fidelity ones for the nursing schools were reportedly 45.2-71.0, 86.5, and 24.2%, respectively [6].

The limitation of this study is that the data are based on a literature review, rather than on an actual condition survey. Furthermore, investigations to elucidate the history and the current status of this subject should be required, and published in English.

We conclude that, in Japan, much more importance should be attached to simulation-based undergraduate nursing education as a teaching methodology, as well as a research field. The optimal socioeconomic environment to introduce sophisticated patient simulators must be prepared by the government.

REFERENCES

[1] B. Ricketts, "The role of simulation for learning within pre-registration nursing education - a literature review," Nurse Educ. Today, vol. 31, pp. 650-654, Oct. 2011, doi: 10.1016/j.nedt.2010.10.029.

[2] T. Ryall, B. K. Judd, and C. J. Gordon, "Simulation-based assessments in health professional education: a systematic review," J. Multidiscip. Healthc, vol. 22, pp. 69-82, Feb. 2016, doi: 10.2147/JMDH.S92695.

[3] G. Alinier G, "A typology of educationally focused medical simulation tools," Med. Teach, vol. 29, pp. 243-250, October 2007.

[4] G. Alinier, B. Hunt, R. Gordon, and C. Harwood, "Effectiveness of intermediate-fidelity simulation training technology in undergraduate nursing education," J. Adv. Nurs, vol. 54, pp. 359-39, May 2006, doi: 10.1080/01421590701551185.

[5] A. Walker, C. Earl, B. Costa, and L. Cuddihy, "Graduate nurses' transition and integration into the workplace: a qualitative comparison of graduate nurses' and Nurse Unit Managers' perspectives," Nurse Educ. Today, vol. 33, pp. 291-296, Mar. 2013, doi : 10.1016/j.nedt.2012.06.005.

[6] N. Kuroda, and Y. Orii, Kangokisokyoiku ni okeru shumiieeta wo mochiita shimumireeshon kyiku no jittaichosa [Acutual condition survey of simulation-based undergraduate education in Japan]," J Jap. Assoc. Sim. Educ. in Healthc. Prof, vol. 4, pp. 22-28, Jun. 2016.

[7] R. P. Cant, and S. J. Cooper, "Use of simulation-based learning in undergraduate nurse education: An umbrella systematic review," Nurse Educ. Today, vol. 49, pp 63-71, Feb. 2017, doi: 10.1016/j.nedt.2016.11.015.

Effects of a Blended Learning System on Improving the Physical Assessment Ability of Nursing Students

Machiko Saeki Yagi
School of Nursing
Jichi Medical University
Shimotsuke, Japan
email: mati-s@umin.ac.jp

Natsuko Miura
School of Nursing
Iwate Prefectural University
Takizawa, Japan
email: natsuko@iwate-pu.ac.jp

Shigeki Tsuzuku
Graduate School of Instructional Systems
Kumamoto University
Kumamoto, Japan
email: tsuzuku@kumamoto-u.ac.jp

***Abstract*—This study assessed the effects of a blended learning system—combining e-learning with simulation training—on improving the physical assessment ability of nursing students in Japan. Specifically, we evaluated how e-learning improved physical assessment knowledge and its application in simulation training. We also reviewed their evaluations of the blended learning system, comprising e-learning and simulation training, after they experienced the system and the influence of blended learning on their desire to learn. We used an interventional study design with a questionnaire survey. The results indicated that e-learning improved participants' ability of checking vital signs and information gathering skills. However, e-learning as regards conducting auscultation, and palpation showed no significant learning effects and simulation training helped to do auscultation and palpation. Additionally, e-learning helped participants recognize the relevance of learning content to their professional practice. In conclusion, using e-learning to study some physical assessment skills was effective and it's necessary to combine e-learning and simulation to improve physical assessment ability as blended learning.**

Keywords-e-learning; nursing students; physical assessment; nurse education.

I. INTRODUCTION

Nurses are increasingly expected to obtain skills related to grasping their patients' situations, particularly the ability to provide advanced and accurate physical assessments [1]. Thus, nurses must begin physical assessment training that is suitably adapted to clinical practice (e.g., simulation education) as early as possible, ideally in their basic nursing education. As physical assessment relies on nurses appropriately integrating knowledge and technique, such as conducting an accurate physical examination and judging a situation based on available information, learning to perform it is challenging [2]. Optimal methods for teaching integrative content to large groups of people thereby need to be examined to improve physical assessment education.

One potentially effective method of ensuring optimal learning for each individual in group learning situations is e-learning. In the context of continuing education for physicians, e-learning has been demonstrated to be as effective as conventional teacher-led education [3]. In other words, e-learning appears to be an effective method of instruction, as it does not limit the learning environment or time for learning, at least for knowledge-level content. In recent years, the breadth of the utilization of e-learning in medical education has been expanding, marking a shift from mere passive knowledge transmission to active, self-motivated learning [4]. Indeed, self-motivated learning using e-learning tools has become a critical feature of nursing education. Hence, learning that effectively utilizes ICT needs to be implemented from early on in nurses' education. Current students receiving basic nursing education are known as "digital natives"; they are from a generation for whom the Internet has existed since birth [5], indicating that their information literacy is higher than that of past generations. Thus, they have a high likelihood of readily taking to e-learning methods in their education.

Although e-learning affords numerous benefits to learners, it also presents a number of challenges. First, e-learning relies on learners' ability to self-manage their learning process, and second, e-learning methods must possess features that help maintain the desire to learn [6]. For example, massive open online courses (MOOCs) [7] can theoretically provide instruction to an extremely large number of participants, but they do not allow learners to self-manage their learning, and students tend to drop out of these programs owing to diminished motivation to learn. One strategy that has been proposed to resolve these problems in MOOCs is "blended learning," which combines face-to-face and e-learning methods. We thus hypothesize that such blended learning training may help overcome the challenges of e-learning. This study aimed to investigate the effects and influence on motivation of a blended learning system to improve physical assessment ability. The remainder of this paper is organized as follows. Section II presents the methods. Section III provides an overview of our results. Section IV discusses implications and limitations. Finally, conclusions and future work are detailed in Section V.

II. METHODS

A. *Participants and Data Collection Methods*

Of the 82 junior nursing students who had already completed physical assessment training at A university, we recruited the 34 students who consented to participate (33 females and 1 male aged 20 to 22 years). A preliminary testing of the participants' computer literacy revealed that they were able to connect to the Internet and could effectively use mobile devices (e.g., laptop, smartphones), but had not had previous e-learning experience. Additionally, they either did not have any experience with simulation training or had undergone it once or twice. Data collection was conducted through objective assessment (i.e., correct answer rate on an e-learning test, and a checklist evaluating simulation results) and subjective evaluation of the e-learning experience (questionnaire). The investigation period was from September to October 2012.

B. Design of the Physical Assessment Training Using E-learning and Simulation

1) Learning Objective

The learning objective of the physical assessment training was "able to visit the patient's bedside and comprehend their condition." Accordingly, we created a checklist of the skills that participants should be able to perform at the end of the training, including visual inspection, performing a medical interview, measurement of vital signs, and performing auscultation and palpation. We created a simulation scenario wherein they carried out a physical assessment of a breast cancer patient during chemotherapy. This blended learning was designed to create suitable educational materials to help nursing students learn real-world clinical situations, by using stories to help them better anticipate the realities of clinical practice. The characteristic points of the materials were the use of Japan's national nursing examination, modified using scenarios and actual clinical cases, and story-type e-learning that included learning opportunities for assessments, clinical thinking, patient safety, and communication.

2) E-learning Design

An open-source learning management system, Moodle [8], was used to design the e-learning system. Specifically, the e-learning system presented patients on a screen, on which participants were supposed to perform a physical assessment using the displayed data (Fig 1). The system also included a test on content related to physical assessment. The test was conducted before the simulation training to confirm if the participants possessed the knowledge necessary to take part in the simulation. As in previous research [9], the structure of the e-learning system involved solving test problems within the same account as the simulation, thus serving as an introduction to the simulation. The e-learning covered the period of the student's entry to the hospital room, conduct of the physical assessment, and exit from the room. It was configured such that participants answered 14 multiple choice questions and 2 short-answer questions, for a total of 16 questions.

3) Simulation Design

The simulation was conducted in a simulated hospital room (Fig 2). An advanced simulator that enables the measurement of patients' blood pressure, pulse, and respiratory rate was used. During the simulation, learning achievement was evaluated using the simulation checklist mentioned previously. In the medical interview, the patient's vocal responses were produced by an instructor with a microphone.

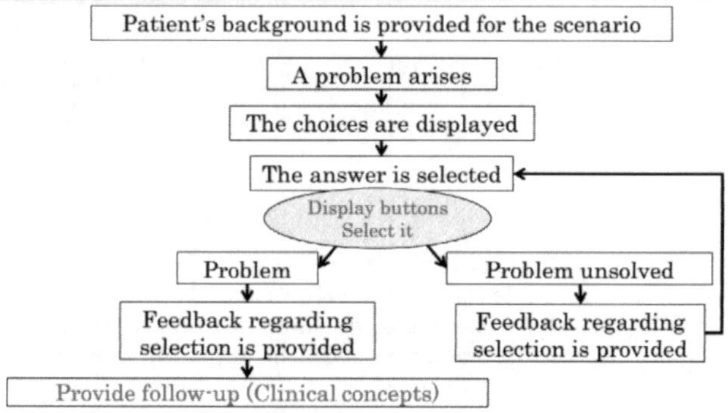

Figure 1. Flow of e-learning

C. Questionnaire Creation

The questionnaire assessed participants' opinions of the e-learning system (including amount of learning material, difficulty of questions, visibility, and ease of operation) and its effect on their desire to learn using a five-point Likert scale (extremely dissatisfied, slightly unsatisfied, neither satisfied nor dissatisfied, slightly satisfied, and extremely satisfied) and a free-comment section.

Desire to learn and the influencing factors for the successful implementation of learning were assessed using the four-subscale Instructional Materials Motivation Survey (IMMS), which was created based on the ARCS model [10]. This model contains four components: attention, relevance, confidence, and satisfaction. The IMMS, comprising 36 items, was designed to measure individuals' reactions to self-driven teaching materials, such as e-learning. A five-point Likert scale was used to rate each item, with higher scores indicating a more positive opinion. The IMMS includes items expressed negatively, which are reverse scored. In other words, before adding these answers to the total score, items given a rating of 5 were converted to have a rating of 1, with 4 being converted to 2, 3 remaining the same, 2 to 4, and 1 to 5.

After completing the entire training sequence, participants who gave their consent were interviewed to explore their thoughts on the e-learning and simulation, and to determine the influence of the simulation training on learning effectiveness and desire to learn.

D. Statistical evaluation

For the e-learning, test scores, times, and times of simulation are expressed as means ± standard deviations. For the questionnaire, scores are expressed as means and median values (quartile range). The free-response descriptions were listed as written by participants. All statistical analyses were performed using IBM SPSS Statistics 22 (IBM Corp., Armonk, NY). A significance level of 5% was used in the analysis.

E. *Ethical Considerations*

This study was conducted with the approval of the medical ethics committees of Nagoya University, Japan; the survey was approved in June 2013 (No. 2013-0049).

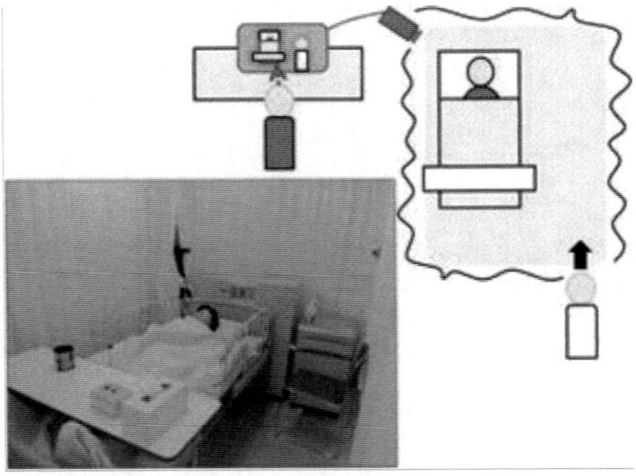

Figure 2. Simulated hospital room: The instructor watches participants' performance via video to ensure a safe learning environment.

III. RESULTS

A. *Research Participation*

Of the participants, 34 finished the e-learning and simulation, whereas 25 completed the questionnaire. Interview data were obtained from four participants; most of the participants had to attend clinical training.

B. *E-learning*

The correct answer rate on the test was 91.1% ± 8.5%: that for the measuring site of temperature and pulse, and the side effect of chemotherapy were for each 73.8%, 68.5%, 73.8%, less than 75%. Regarding the time taken to complete the test, the participants had a mean time of 21 min 49 s ± 9 min 15.

C. *Simulation*

According to the checklist analysis results, all 25 participants were able to greet the patient upon entering room; measure the patient's pulse, blood pressure, respiratory rate, and temperature, and then interpret the measured values; and exit the room after a polite goodbye. Confirmation of food intake, sleeping condition, excretion condition, and side effect of chemotherapy was carried out by 28, 21, 25, and 30 individuals, respectively; only two participants used information obtained from touching, such as auscultation and palpation. The mean completion time of the simulation was 10 min 32 s ± 2 min 12. After the feedback about simulation from the instructor, all participants understand the needs of auscultation and palpation and tried to examine by auscultation and palpation once again.

D. *Questionnaire and Interview Results*

1) *Evaluation of E-learning and Simulation*

The results of the questionnaire assessing the e-learning characteristics are shown in Table 1. The participants chose item 3 in the scale as their response for all of the questionnaire items. From the free-response answers, one participant reported that the number of questions and difficulty of the e-learning was "exactly the right number and difficulty." Another reported that it was "a little too much to do in my spare time." Regarding visibility and operability, which had the most free-response answers, participants listed various concrete problems, such as "difficult to see," "difficult to use on a cellphone," "froze mid-answer," and "difficult to enter sentences on a smart phone."

According to the interviews, participants reported the following on the simulation: "It was interesting, so I would like to use this method in classes as well"; "Because it was done in the e-learning, I could imagine it"; and "Because I knew what was going to be implemented beforehand [owing to the e-learning], I was able to do it." One participant reported, "Although I understood the content, I became nervous and forgot it [during the simulation]." Additionally, participants reported contrasting views, such as "I was able to reconfirm my knowledge with the test" and "It was unnecessary because I was able to understand from the feedback during the simulation."

TABLE I. THE RESULTS OF THE QUESTIONNAIRE ASSESSING THE E-LEARNING CHARACTERISTICS

Contents	Mean, Median; Quartile range
Number of questions	2.9, 3; 3-3
Time taken	2.9, 3; 3-3
Difficulty	2.8, 3; 3-3
Readability	3.3, 3; 3-4
Character size	3.4, 3; 3-4
Screen design	3.7, 3; 3-3
Operability	3.0, 3; 3-3

2) *Influence on Desire to Learn*

Table 2 shows an IMMS item. Items 1 "When I first looked at this lesson, I had the impression that it would be easy for me" and 9 "There were stories, pictures, or examples that showed me how this material could be important to some people" (included in the Confidence and Relevance subscale), had a low median (3 points). According to the free-response answers, a number of participants expressed wanting more developmental content included in the e-learning. The participants' comments

TABLE II. THE RESULTS OF INSTRUCTIONAL MATERIALS MOTIVATION SURVEY

Items	Contents	ARCS	reverse	Mean, Median; Quartile
2	There was something interesting at the beginning of this lesson that got my attention.	A		3.5, 4; 4-4
8	These materials are eye-catching.	A		3.7, 3; 3-5
11	The quality of the writing helped to hold my attention.	A		3.5, 3; 3-4
12	This lesson is so abstract that it was hard to keep my attention on it.	A	○	4.2, 4; 4-5
15	The pages of this lesson look dry and unappealing.	A	○	4.5, 5; 5-5
17	The way the information is arranged on the pages helped keep my attention.	A		3.7, 3; 3-4
20	This lesson has things that stimulated my curiosity.	A		4.1, 4; 4-5
22	The amount of repetition in this lesson caused me to get bored sometimes.	A	○	4.1, 4; 4-5
24	I learned some things that were surprising or unexpected.	A		3.6, 4; 4-4
28	The variety of reading passages, exercises, illustrations, etc., helped keep my attention on the lesson.	A		3.6, 4; 4-4
29	The style of writing is boring.	A	○	4.1, 4; 4-5
31	There are so many words on each page that it is irritating.	A	○	4.4, 5; 5-5
6	It is clear to me how the content of this material is related to things I already know.	R		3.9, 4; 4-4
9	There were stories, pictures, or examples that showed me how this material could be important to some people.	R		2.9, 3; 3-3
10	Completing this lesson successfully was important to me.	R		4.3, 4; 4-5
16	The content of this material is relevant to my interests.	R		4.0, 4; 4-4
18	There are explanations or examples of how people use the knowledge in this lesson.	R		4.1, 4; 4-5
23	The content and style of writing in this lesson convey the impression that its content is worth knowing.	R		3.6, 4; 4-4
26	This lesson was not relevant to my needs	R	○	4.2, 4; 4-5
	because I already knew most of it.			
30	I could relate the content of this lesson to things I have seen, done, or thought about in my own life.	R		4.1, 4; 4-5
33	The content of this lesson will be useful to me.	R		4.6, 5; 5-5
1	When I first looked at this lesson, I had the impression that it would be easy for me.	C		2.8, 3; 3-3
3	This material was more difficult to understand than I would like for it to be.	C	○	4.1, 4; 4-5
4	After reading the introductory information, I felt confident that I knew what I was supposed to learn from this lesson.	C		3.4, 3; 3-4
7	Many of the pages had so much information that it was hard to pick out and remember the important points.	C	○	4.1, 4; 4-5
13	As I worked on this lesson, I was confident that I could learn the content.	C		3.3, 3; 3-4
19	The exercises in this lesson were too difficult.	C	○	3.3, 3; 3-4
25	After working on this lesson for awhile, I was confident that I would be able to pass a test on it.	C		3.3, 3; 3-4
34	I could not really understand quite a bit of the material in this lesson.	C	○	4.6, 5; 5-5
35	The good organization of the content helped me be confident that I would learn this material.	C		4.0, 4; 4-4
5	Completing the exercises in this lesson gave me a satisfying feeling of accomplish¬ment.	S		4.0, 4; 4-4
14	I enjoyed this lesson so much that I would like to know more about this topic.	S		3.8, 4; 4-4
21	I really enjoyed studying this lesson.	S		3.9, 4; 4-4
27	The wording of feedback after the exercises, or of other comments in this lesson, helped me feel rewarded for my effort.	S		3.9, 4; 4-4
32	It felt good to successfully complete this lesson.	S	○	4.0, 4; 4-4
36	It was a pleasure to work on such a well-designed lesson.	S		4.1, 4; 4-5

suggested a desire to use the e-learning system in the future, such as "In practice, I felt like I would be able to use it"; "I want to do normal exercises as well as e-learning and simulation"; and "As long as you know how it works, I think you could practice simulation similar to this with a classmate."

IV. DISCUSSION

A. *Effects of Blended Learning*

Although simulation is an effective learning method and typically provides a greater sense of understanding, it often causes students to experience significant mental strain, thus making them require learning support. We developed a pre-simulation e-learning system to provide such support. The results showed that, after using the e-learning system, all of the participants performed the greeting, vitals measurement, and complete physical assessment based on the gathered measurements, even if it was their first experience with the simulation. This success can be attributed to participants' completion of the skill building and mock training exercises via e-learning system before the simulation. In other words, the e-learning program enabled them to imagine the content of the simulation and thus integrate relevant knowledge and skills into their simulation experience.

However, participants did not completely learn to perform auscultation and palpation, according to information obtained from medical interviews [11]. Therefore, the e-learning items related to this content may need to be modified. Specifically, the e-learning system can be modified such that, when repeatedly used, items that have been completely learned are excluded, thereby enabling users to focus on items that were not completely learned. Such a modification would allow for the delivery of deliberate and effective training aligned with the learning objectives of nurse education, referred to as deliberate practice [12][13].

Learning self-management has been identified as a key factor for continued education via e-learning [6]. To succeed in remote self-learning, participants may need to receive adequate explanation on the necessity of post-training tests in acquiring knowledge related to medical interviews, auscultation, and palpation, which were areas that remained challenging after the simulation training.

B. *Effects of E-learning and Training on Motivation to Learn*

The ARCS model is a method of organizing the factors that influence motivation to learn [14], described by Suzuki as follows: "A) First, attention is captured. 'This looks interesting—there's something to this.' R) Next, one realizes the relationship to themselves; knowing the learning task and realizing that 'it looks rewarding and it is related to my values.' Not only is the future value of the task significant, but also enjoying the learning process is valued. However, even if one finds significance in their learning, one can lose motivation when one recognizes that there is low possibility to accomplish the learning goal. C) On the contrary, if one has successful experiences in the first learning stages and can associate the experience with the endeavor, leading to the perception that 'I can manage it,' confidence is facilitated. S) If one can feel fulfillment after looking back on the learning process and its accomplishment, it then leads to motivation to learn." In the present study, the participants recognized e-learning as "having an important relationship to themselves," and they utilized this recognition as motivation to learn. To improve learners' confidence, we propose the following strategies: 1) Share evaluation criteria and allow learners to tackle tasks with a prediction of their possibility for success in mind; 2) adjust the difficulty of the system so that they can have a meaningful and successful experience; and 3) provide opportunities and feedback that regulate learning, thus encouraging them to be aware that they can achieve their goal by themselves [15]. A number of participants commented that the training could feasibly be implemented as peer training and skill review. Repetitive practice through sharing stories with peers and skill review has been shown to help learners acquire self-confidence [16]. Further consideration of task difficulty will also facilitate continuous learning in this context.

C. *Determinants of Successful E-learning*

As described above, e-learning relies on independent learning, which means that it has an inherent risk of dropout. We thus set learning objectives to help foster learning motivation and created teaching materials relevant to real-world clinical settings. In this study, the item "The content of this lesson will be useful to me" was rated 4.6, 5; 5-5 point, indicating that blended learning sustained participants' motivation. Additionally, the item "I could not really understand quite a bit of the material in this lesson" was reverse scored, also totaling 4.6, 5; 5-5 point. The difficulty of blended learning and acquisition of confidence through simulation may be factors that influenced their motivation. In this respect, the structure of the blended learning needs to be improved such that it can help participants gain confidence in their ability to perform the learning exercises and promote their self-learning through repetitive practice and detailed explanations for items identified as their weakness.

Operability is another factor that may have influenced participation. We expected this tendency in designing the system, and had tested the teaching material on smartphones prior to the survey. However, our interview results showed that when smartphones were used, the participants reported longer working time and poorer operability. This finding coincides with previous reports that the physical environment for online learning, such as Internet and terminal devices, influences compliance with e-learning [17].

For example, poor video teaching materials can reduce the motivation to learn [18]. Future research is needed to investigate additional factors related to operability, the differential effects of computers and smartphones, and the effects of the tendency to use smartphones for mobile learning, which is common among nursing students.

Recently, an interactive digital simulator for problem solving and clinical reasoning, both of which require physical assessment skills, was developed in Europe [19]. The digital simulator was developed for medical students and clinicians and uses virtual patients. Problem solving and clinical reasoning are needed in clinical nursing practice [20]. Future studies are needed to develop learning materials for nurses to acquire these required competencies immediately.

V. CONCLUSION

The results of the present study demonstrate that the e-learning system we developed was partially effective for improving physical assessment ability like checking vital signs and information gathering skills. However, our results show that e-learning is insufficient for augmenting the ability of auscultation and palpation, suggesting that teaching materials related to these skills must be improved using simulation training. Additionally, e-learning helped participants recognize the relevance of learning content to their professional practice. In conclusion, it's necessary to combine e-learning and simulation to improve physical assessment ability as blended learning.

The advancement of remote education provides students with the opportunity to learn anywhere as opposed to strictly in classrooms. E-learning systems allow individuals attempting to learn medical practice to learn effectively and efficiently in their own environment. Thus, we aim to continue improving the current e-learning system, by making it more operable, to ensure that it sustains motivation to learn, as such, encourages continuous learning.

ACKNOWLEDGMENT

Part of this research was supported by Grant-in-Aid for Young Scientists (B), 25862108, from Japan Society for the Promotion of Science (JSPS), JAPAN.

REFERENCES

[1] J. K. Hayden, R. A. Smiley, M. Alexander, S. Kardong-Edgren, and P. R. Jeffries, "The NCSBN National Simulation Study: A longitudinal, randomized, controlled study replacing clinical hours with simulation in prelicensure nursing education," J Nurs Regul, vol. 5, no. 2, pp. S1–S64, 2014.

[2] L. Bickley and P. G. Szilagyi, Bates' guide to physical examination and history-taking. Philadelphia, Pennsylvania: Lippincott Williams & Wilkins, 2012.

[3] M. Fordis, J. E. King, C. M. Ballantyne, P. H. Jones, K. H. Schneider, S. J. Spann, et al., "Comparison of the instructional efficacy of Internet-based CME with live interactive CME workshops: a randomized controlled trial," JAMA, vol. 294, no. 9, pp. 1043–1051, 2005.

[4] J. G. Ruiz, M. J. Mintzer, and R. M. Leipzig, "The impact of e-learning in medical education," Acad Med, vol. 81, no. 3, pp. 207–212, 2006.

[5] C. Jones, R. Ramanau, S. Cross, and G. Healing, "Net generation or Digital Natives: Is there a distinct new generation entering university?," Comput Educ, vol. 54, no 3, pp. 722–732, 2010.

[6] J. Keller and K. Suzuki, "Learner motivation and e-learning design: A multinationally validated process," J Educ Media, vol. 29, no. 3, pp. 229–239, 2004.

[7] J. Daniel, "Making sense of MOOCs: Musings in a maze of myth, paradox and possibility," J Interac Media Educ, vol. 2012, no. 3, 2012.

[8] M. Dougiamas and P. Taylor, "Moodle: Using learning communities to create an open source course management system," World Conference on Educational Media and Technology, vol. 1, pp. 171–178, 2003.

[9] M. S. Yagi, K. Ikegami, Y. Shibata, S. Shiba, and M. Miyoshi, "The design of the story-based learning materials to imagine realistic clinical practice using goal-based scenario theory," Japan Journal of Health Professional Development, vol. 3, no. 2, pp. 69–76, 2015.

[10] J. M. Keller, Motivational design for learning and performance: The ARCS model approach. New York City: Springer Science & Business Media, 2009.

[11] J. H. Barsuk, W. C.McGaghie, E. R.Cohen, K. J.O'Leary, and D. B. Wayne, "Simulation-based mastery learning reduces complications during central venous catheter insertion in a medical intensive care unit," Crit Care Med, vol. 37, no. 10, pp. 2697–2701, 2009.

[12] K. A. Ericsson, K. Nandagopal, and R. W. Roring, "Toward a science of exceptional achievement," Annals of the New York Academy of Sciences, vol. 1172, no. 1, pp. 199–217, 2009.

[13] M. H. Oermann, S. Kardong-Edgren, T. Odom-Maryon, B. F Hallmark, D. Hurd, N. Rogers, et al., "Deliberate practice of motor skills in nursing education: CPR as exemplar," Nurs Educ Perspec, vol. 32, no. 5, pp. 311–315, 2011.

[14] J. M. Keller, Motivational design for learning and performance: The ARCS model approach. New York City: Springer Science & Business Media, 2009.

[15] K. Suzuki, "On the Framework of Designing and Developing 'appealing instruction': The ARCS Motivation Model," The Japan Association for Educational Media Study, vol. 1, no. 1, pp. 50–61, 1995.

[16] K. M. Lundberg, "Promoting self-confidence in clinical nursing students." Nurse Educ, vol. 33, no. 2, pp. 86–89, 2008.

[17] C. Power, H. Petrie, V. Sakharov, and D. Swallow, "Virtual learning environments: another barrier to blended and e-learning," in International Conference on Computers for Handicapped Persons, Berlin and Heidelberg: Springer, pp. 519–526, 2010.

[18] M. S. Yagi, N. Kumagai, C. Funakoshi, and M. Masuda, "Development of e-learning material for sophomore nursing students to simulate clinical practice," The Association for Medical Education in Europe, 270, #4JJ16 (126905), 2016.

[19] Body Interact. http://bodyinteract.com/ [retrieved: Feburary, 2017]

[20] P. E. Benner, P. L. Hooper-Kyriakidis, and D. Stannard, Clinical wisdom and interventions in acute and critical care: A thinking-in-action approach. New York City: Springer Publishing Company, pp. 10–11, 2011.

Development of Educational Medical Equipment Capable of Displaying Abnormal Value
—To Turn Simulated Patient into Real Patient—

Natsuko Miura, Kazuma Takahashi
Faculty of Nursing
Iwate Prefectural University
Takizawa, Japan
e-mail: natsuko@iwate-pu.ac.jp, t_kazuma@iwate-pu.ac.jp

Takanori Ito
Iwate Monozukuri Software Integration Technology
Iwate Prefectural University
Takizawa, Japan
e-mail: itou_t@ipu-office.iwate-pu.ac.jp

Abstract— **In order to acquire clinical practice capability such as nursing, such a learning is required that assumes a situation similar to clinical setting. Therefore, simulation education has been increasingly performed by using simulator and simulated patients. Even though highly functional simulator is capable of displaying changes of vital signs on a monitor, any educational device to measure biological information, such as blood-pressure manometer, thermometer and pulse oximeter, has not been developed yet. As educational medical equipment capable of intentionally displaying abnormal value has been developed in the study, we would like to introduce the equipment.**

Keywords-simulation; Physical Assessment; Simulated Patient; Clinical reasoning.

I. INTRODUCTION

We developed the equipment which could display an abnormal value intentionally to practice measurement of vital signs, we would like to report on the results.

Along with a rapid progress of low birth rate and an aging population as well as advanced medical treatment in Japan, scope and opportunities for nursing students to practice nursing techniques by internship at hospital tend to be limited due to people's enhanced consciousness for medical care safety [1]. Amid such situation, a nursing education based on simulation equipment is going to be disseminated widely that allows students to repeatedly experience various practical scenes without impairing patients' safety and provides learner-oriented education, and the equipment has been introduced to our university [2].

In section 2, we reported simulation in the nursing education, and the problem. In section 3, we explained equipment for the education that we developed.

II. SIMULATION EDUCATION IMPLEMENTED IN OUR UNIVERSITY, AND THE PROBLEM

There are three types of trainings conducted in the simulation education, i.e., task training for acquiring techniques, algorithm training like learning of Basic Life Support, and situation-based training in which various clinical situations are reproduced [3].

In our university, a skill test of vital signs measurement for their first year and the situation-based training is performed for their second year for patient assessment using simulated patient prior to clinical training. Visiting simulated patients to measure their vital signs by using the skill learned in the first year, conditions of patients are assessed in the training adding information obtained through conversation with the patients. In their third year, students acquire an observation skill in the area of adult nursing science for general condition of patients including their vital signs after gastrectomy.

In a conventional situation-based training, our university contrived to present automatic slide show in Power Point for reproducing fluctuation in addition to display of a card with written numerical values and verbal suggestion by teachers (Figure1). However, these efforts had problems to

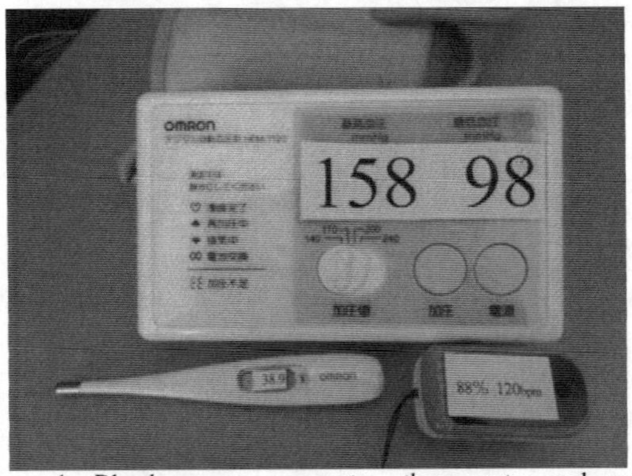

Figure 1. Blood-pressure manometer, thermometer and pulse oximeter, which we put paper on to reproduce an abnormal value.

discourage students' concentration and suspend actions due to the situation different from reality.

Education with a highly functional simulator capable of controlling vital signs, breath sound and heart sound by a computer has been practiced in areas such as intensive care [3] and operation room [4].

In addition, simulated patients who have been trained to play patient-specific sentiment and personality not only for clinical history and physical findings as much as possible have been widely used for medical staff education in addition to co-medical students [3]. The highly functional simulator is unable to make conversation, however, it is said to have limitation to improve communication capability since a person speaks through a microphone in case of a scenario that requires conversation [5]. Further, since normal values are displayed when vital signs are measured because simulated patients are healthy people it is required to show abnormal values with a card [6] or verbally at a designated timing [2].

The problem in such setting is that nursing in a course of natural flow cannot be reproduced.

III. DEVELOPMENT OF EDUCATIONAL VITAL SIGN MEASURING EQUIPMENT

The equipment newly developed this time is possible not only to set up numerical values arbitrarily but also to add and subtract any numerical value from setup value in order to express fluctuation of vital signs (Figure 2). It works in such a way that a value near to setup value is displayed for each measuring by reducing the range of the addition and subtraction and normal value or abnormal value is displayed by increasing the range.

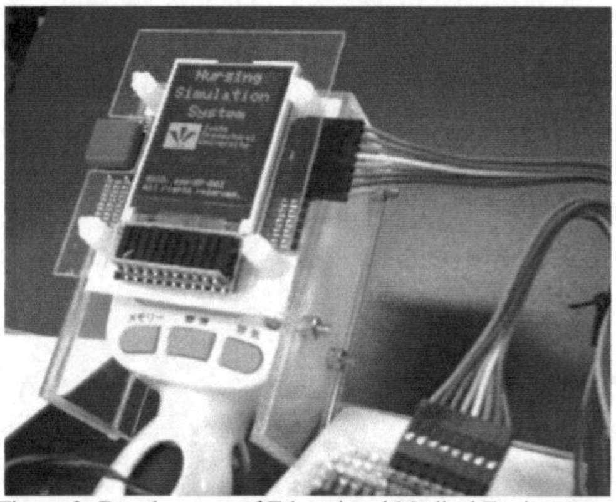

Figure 2. Development of Educational Medical Equipment.

The ability that a student can get with a new equipment;

- Ability for assessment of vital signs
- Ability for clinical reasoning
- Ability to consider for the patients
- Ability to cope with abnormal case

In case of simulation of repeated support for patients presenting with fever, for example, range of adjustment is set up smaller to show higher value at each measurement of body temperature. In case of measuring blood pressure of patients with a complaint of nausea, as value to be displayed becomes different each time of measurement by setting the range of adjustment larger, it may be applied to training of clinical inference.

Further, time required until display may be set up arbitrarily by seconds. For example, it is possible to reproduce time required until displaying measured value of actual and predict types of measurement thermometer with different measuring style and it may be also possible either to set up the time for actual value of measuring device to be displayed or to set up it shorter aiming at reducing the time to be taken for the scenario depending on the purpose of simulation education.

Even though the procedures mentioned above are performed of course by examination on paper-patient or presentation by a card, it is possible to reproduce circumstance more similar to clinical front by measuring vital signs while making conversation with simulated patient. As it is also possible to set up values to be displayed and time in a more detailed manner depending on contents of the scenario of simulation education, readiness of students and capability of teachers, we believe it is useful.

Further, the equipment is capable of sending setup data to a server by wireless. It is not required to set up each device since subsequently multiple devices such as blood-pressure manometer, thermometer and pulse oximeter may respectively access to a parameter setting server by an interval of three seconds during waiting. It becomes also possible to perform simulation with the same content in multiple booths at the same time.

We believe it is useful because it results in saving the effort of a person in charge of training preparation. Even though it is only at prototype stage as of now, we hear such impressions from teachers who have practiced the simulation education at our university that they are interested and want to use it.

In the future, we have planned to clarify the educational effect of the newly developed equipment for students by comparing simulation education using the new equipment with conventional method.

IV. CONCLUSION

The equipment we developed strengthens the student's clinical practice ability. Our problem is downsizing.

REFERENCES

[1] Ministry of Health, Labor and Welfare, "Investigative Commission report regarding enhancement of nursing basic education" http://www.mhlw.go.jp/shingi/2007/04/dl/s0420-13.pdf, [retrieved: February, 2017]

[2] Y. Endo, N. Miura, H. Chieda, T. Takeda, "A trial of using simulation in undergraduate nursing education ", Journal of the Faculty of Nursing, Iwate Prefectural University, 16, pp.43-46, 2014.

[3] Y. Abe, "Simulation education in medical care", Journal of the Japanese Society of Intensive Care Medicine, 23, pp.13-20, 2016. .

[4] R. Takashiki, K. Abe, C. Saito, T. Horiguchi," Effect of simulation education for operation department nurses to cope with cases of intubation difficulty". Journal of Japanese Association for Operating Room Technology, 37 (2) , pp.39-41, 2016.

[5] Ministry of Health, Labor and Welfare, "Investigative Commission report regarding contents and methods of nursingeducation", http://www.mhlw.go.jp/stf/houdou/2r98520000013l0q-att/2r98520000013l4m.pdf, [retrieved: February, 2017]

[6] M. Nogi, "Temperature check simulation to strengthen thinking ability of new recruits", Nursing, pp. 88-92, 2013.

Demonstration of The KINECT-Based Auscultation Practice System

Yoshitoshi Murata, Kazuhiro Yoshida
Faculty of Software and Information Science
Iwate Prefectural University
Takizawa, Japan
e-Mail: y-murata@iwate-pu.ac.jp, kyoshida@ipu-office.iwate-pu.ac.jp

Natsuko Miura
Faculty of Nursing
Iwate Prefectural University
Takizawa, Japan
e-Mail: natsuko@iwate-pu.ac.jp

***Abstract*— Students in medical and nursing schools must practice auscultation. Students usually learn disease sounds, correct points and order for locating a stethoscope on a body in the practice. Humanoid-type simulators have been widely introduced to practice auscultation. However, most humanoid-type simulators cannot detect whether a stethoscope is located on a body, and on which part of body a stethoscope is placed. Since they are too expensive, the number of them is not enough for the number of students in a class. We developed a low-cost and high performance system for the practice of auscultation. In this system, students themselves play the role of a patient, instead of a humanoid, and stethoscope locations on the body are measured with KINECT. Also, appropriate disease sounds including normal ones can be assigned at some points on the upper body. Students listen for such sounds, synchronized with breathing movements, through an earphone or speaker when a stethoscope is placed on assigned points. We plan to conduct a demonstration in which participants will practice auscultation with the system after it is explained to them.**

Keywords-simulator; auscultation; physical assessment; demonstration.

I. INTRODUCTION

Generally, practicing auscultation is a required subject for students in medical and nursing schools. Students usually learn disease sounds, correct points and order for locating a stethoscope on the body in the practice. We proposed a new auscultation practice system to learn auscultation techniques effectively [1]. Humanoid-type simulators [2][3][4] have been widely introduced into medical and nursing schools. These humanoid-type simulators are effective to learn disease sounds. However, it is impossible to detect whether a stethoscope is actually placed correctly on a mannequin. Moreover, correct locations vary among patients according to their body size. Cardionics provides a hybrid simulator in which a student plays the role of a patient instead of a mannequin to solve such problems [5]. In this hybrid simulator, patches are attached on a body to identify correct attachment points.

In our simulator, students themselves are the practice subjects instead of a humanoid model, and the location of a stethoscope can be detected with KINECT, which is a line of motion sensing input devices made by Microsoft [6]. The correct locations are normalized with respect to the positions of both shoulder joints and both hip joints for each student playing the role of a patient. Therefore, our proposed simulator can both show correct locations on a body and detect whether a stethoscope is placed on correct points without patches regardless of the change in body size.

In addition, most existing simulators cannot simulate the timing of breathing or the synchronized forward and backward movements of the upper body. However, our simulator can detect these forward and backward movements of the front body and provide expiration and inspiration sounds synchronized with those movements.

We have developed a prototype system and evaluated it experimentally. The results showed that our system could detect stethoscope placement on a body. Also, our system could detect changes of the front body in breathing.

Moreover, we found that all students could set up our system by themselves. The results of a questionnaire for nursing students and practicing nurses showed that our proposed system was useful for them to learn auscultation.

We plan to conduct a demonstration in which participants will practice auscultation with the system after it is explained to them.

II. SYSTEM CONFIGURATION

As a matter of course, the introduction cost is adequate for the number of students in a nursing school. Among the nursing skills that students have to learn are the recognition of different sounds between different kinds of diseases and the knowledge about placing correct points and order for locating a stethoscope on a body. In the case of respiratory auscultation, students have to listen to respiratory sounds for more than one cycle. Therefore, an auscultation practice system requires the following issues:

- Low cost.
- Simulating real sounds produced due to certain diseases at different points on the body.
- Display correct points for locating a stethoscope.
- Determining whether a stethoscope is located at specified points.
- Determining whether a stethoscope remains on a body for more than one respiratory cycle.

Our practice system is only composed of a PC and a KINECT device, as shown in Fig. 1. Students, instead of mannequins, act as patients to decrease system cost. The stethoscope locations and forward and backward movement of a body during breathing are measured with KINECT. The students can listen for sounds produced due to certain diseases that are generated by the PC through earphones. Therefore, a specific stethoscope and patches are not needed. These also decrease system cost.

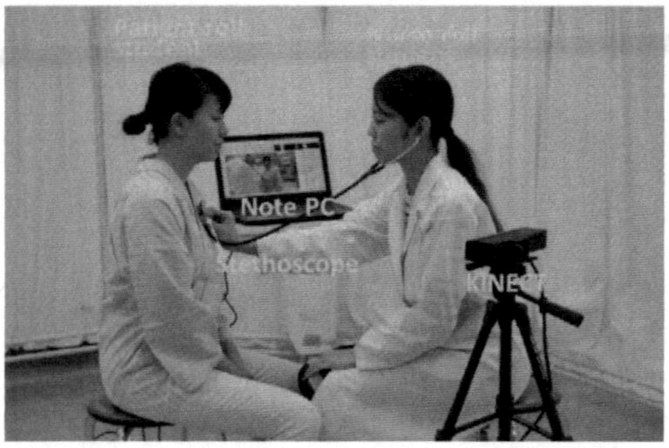

Figure 1. Terminal equipment in our proposed system

III. APPLICATION

We developed a simple application for a teacher to explain practicing auscultation for the lungs:

- Sound assigning program: This program assigns a sound to each area, dividing the upper body into five areas. (See Fig. 2)
- Auscultation teaching program: This program detects on which area a stethoscope should be placed, and it replays a sound assigned to that area. (See Fig. 3)

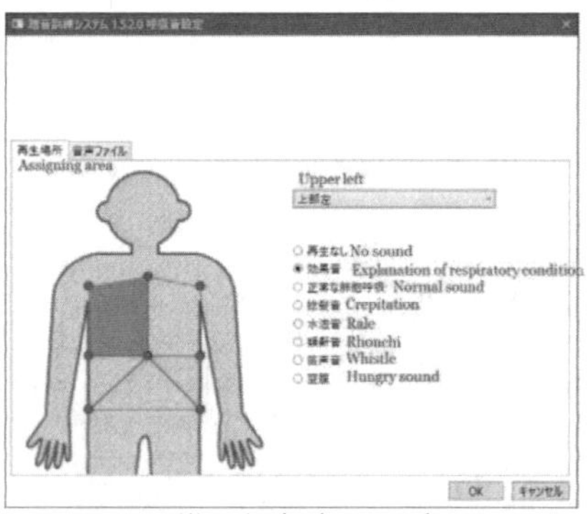

(1) Assigning sound

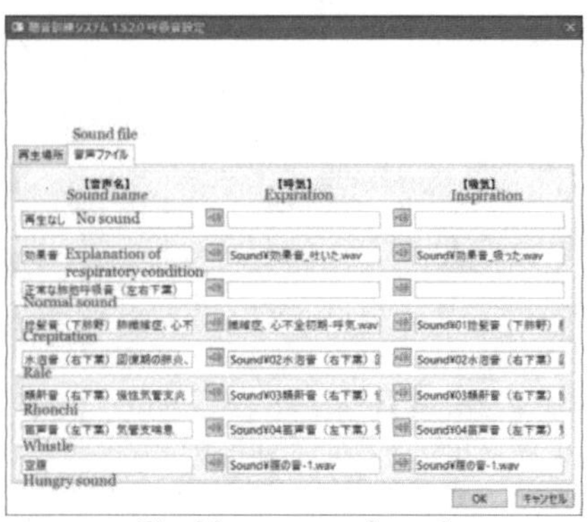

(2) Management of sounds

Figure 2. Operation window to assign sounds to divided body areas

In this application, the upper body is divided to five areas as shown in Fig. 3. Currently eight sounds are provided. A teacher can assign a sound from the above eight sounds to each divided area using the sound-assigning program shown in Fig. 2 (1). Since respiratory sounds are different for expiration and inspiration, an expiration-sound file and inspiration-sound file are assigned for expiration and inspiration using the sound-management program, as shown in Fig. 2 (2).

When a stethoscope is placed on a body of a person playing the role of patient, the area on which a stethoscope is placed is colored to yellow, as shown in Fig. 3. Students can listen for such sounds, synchronized with breathing movements, through an earphone or speaker when a stethoscope is placed on assigned points.

Figure 3. Operation window of the auscultation teaching program

IV. DEMONSTRATION PROGRAM

Since we would like many participants to experience using our simulation system, we plan to demonstrate our auscultation system by

(1) introducing the system configuration,
(2) showing how to set up the system,
(3) explaining how to operate the system,
(4) allowing participants to practice auscultation.

V. CONCLUSION

We believe that our proposed system is less expensive and more useful for practicing auscultation than current humanoid-type simulators. We plan to provide a trial version of this system at no charge for medical and nursing schools to experience our system.

REFERENCES

[1] Y. Murata, N. Miura, and Y. Endo, "Proposal for A KINECT-Based Auscultation Practice System," in Proc. eTELEMED, 2016, pp. 86-91.

[2] Lung Sound Auscultation Trainer "LSAT", Kyoto Kagaku Co., Ltd., https://www.kyotokagaku.com/products/detail02/m81-s.html [retrieved: January, 2017]

[3] Sakamoto Model Corporation, Sakamoto auscultation simulator, http://www.sakamoto-model.com/product/emergency/m164/ [retrieved: January, 2017]

[4] Laerdal, SimMan 3, http://www.laerdal.com/us/SimMan3G [retrieved: January, 2017].

[5] SimScope WiFi (The Hybrid simulator), Cardionics http://www.cardionics.com/simscope [retrieved: January, 2017].

[6] Meet Kinect for Windows, https://developer.microsoft.com/en-us/windows/kinect, [retrieved: January, 2017]

The Use of Dialectical Dialogues in a Personalized Virtual Coach for Obese Emotional Eaters

A Research Protocol for a field study among target group

Aranka Dol[1], Olga Kulyk[2], Hugo Velthuijsen[1], Lisette van Gemert-Pijnen[2], Tatjana van Strien[3]
[1]Institute for Communication, Media & IT, Hanzehogeschool UAS, Groningen
[2]Department of Psychology, Health and Technology, University of Twente, Enschede
[3]Behavioural Science Institute and Institute for Gender Studies, Radboud University, Nijmegen
[3]Department of Health Sciences and the EMGO Institute for Health and Care Research, VU University, Amsterdam
e-mail: [1]{a.dol, h.velthuijsen}@pl.hanze.nl, [2]{o.a.kulyk, j.vangemert-pijnen}@utwente.nl, [3]t.vanstrien@psych.ru.nl

***Abstract*—There is a growing number of eHealth interventions aiming at enhancing lifestyle to address obesity. However, the existing interventions do not take the emotional aspects of obesity into account. Forty percent of the overweight population is an emotional eater. Emotional eaters gain weight because of poor emotion regulation, not just due to bad eating habits. We aim at developing a personalized virtual coach 'Denk je zèlf!' providing support for self-regulation of emotions for young obese emotional eaters. This paper presents an eCoaching model and a research study protocol aiming at the validation of persuasive coaching strategies based on behavior change techniques. Ultimately, we aim at designing a personalized eCoaching framework, allowing us to optimally translate successful behavior change mechanisms and techniques, such as dialectical strategies, into personalized persuasive coaching strategies.**

Keywords-personalized virtual coach; obesity; emotional eaters; persuasive technology; eHealth; mHealth; tailoring messages; coaching strategies.

I. INTRODUCTION

Obesity has become a major societal problem worldwide [1]. The main reason for severe overweight is excessive intake of energy, in relation to the individual needs of a human body. Obesity is associated with poor eating habits and/or a sedentary lifestyle. A significant part of the obese population (40%) overeat due to negative emotions [2]. There is little attention to the personal needs of emotional eaters when it comes to existing health interventions [3]-[5]. Current face-to-face treatments are focused on restructuring cognitions and behavior. There is a need for self-management support and personalized coaching to enhance emotional eaters in recognising and self-regulating their emotions [6][7].

II. RELATED WORK

Virtual coaching systems form a broad and vivid research area [8]-[11]. Since chatbot Eliza was invented by Joseph Weizenbaum [12], many new studies have emerged. Over the last years, virtual coaches have been developed for behavior change support, healthy lifestyle, and physical activity support [13]-[17]. Existing virtual coach applications lack systematic evaluation of coaching strategies and usually function as (tele-)monitoring systems, limited to giving general feedback to the user on achieved goals and/or accomplished (online) assignments [18].

This paper presents an eCoaching model and the research protocol and design rationale for evaluation of persuasive coaching strategies for emotional eaters based on the Dialectical Behavior Therapy and Persuasive System Design model [19].

This research study is a continuation of the previous study on the design and evaluation of a personalized virtual coach for emotional eaters [7]. The long-term goal of the virtual coach is to raise awareness of emotional eaters on their own emotions, and to enhance a positive change of attitude towards accepting the negative emotions they experience.

In the next paragraphs in Section II, we first discuss the theoretical underpinnings of dialectical behavior therapy, persuasive coaching strategies and Ecological Momentary Assessments. Next, we elaborate on the design rationale of a personalized virtual coach. We then present the design of the research protocol in Section III. We finish this paper in Section IV with a discussion on the expected contributions of this research study, conclusion, and future work.

A. *Dialectical Strategies*

Dialectical Behavior Therapy (DBT) focuses on getting more control over one's own emotions by reinforcing skills in mindfulness, emotion regulation, and stress tolerance [37]. Emotion regulation is about recognizing and acknowledging emotions and accepting the fact that they come and go. The behavior change strategies within DBT are based on validation and dialectics [20]. Dialectics changes the users' attitude and behavior by creating incongruence between an attitude and behavior since stimuli or the given information contradict with each other. We believe that the integration of the dialectical behavior change strategies and persuasive features (PSD model) will enhance the personalization of the virtual coach for emotional eaters.

Emotional eaters deal with an invalidating daily life environment. DBT offers powerful "mechanisms of change" with a well-balanced mix of being validated in their perception of negative emotions and a confrontation (dialectical) with a practical focus on changing problem behavior. Our eCoaching model translates these mechanisms into persuasive dialogues and persuasive features.

Validation strategies suggest responding in an empathic way, by hearing another person's point of view and accepting them (and their emotions) without judging. Dialectical strategies focus on confronting the user with a practical focus on changing problem behavior; the key is in finding a balance between acceptance of strong feelings and emotions and change by adapting feelings and emotions using emotion regulation and interpersonal effectiveness.

B. *Persuasive Technology*

Persuasive Technology (PT) is a significant predictor for adherence and offers strategies to reinforce the validation – and dialectical strategies' influence on attitudes and behaviors. We believe that PT can assist emotional eaters in attitude change and acceptance of their own emotions. It is essential to identify the intended outcome or change of the intervention, prior to determining the design principles and coaching strategies that would positively contribute to the persuasiveness in any way. We chose Persuasive Systems Design (PSD) model [19] for translating behavior change techniques from DBT into persuasive coaching strategies, since PSD focuses on persuasion context and easily implementable persuasive design features.

C. *Ecological Momentary Assessments*

Ecological Momentary Assessment (EMA) is applied to systematically collect data about daily experience and feelings of the users, as well as the context of use, for instance, through a user-friendly smartphone application. Experience sampling is widely used in daily psychiatry practice [21]-[22], nowadays allowing to real-time measure and record encountered feelings and emotions through mHealth applications.

D. *Design rationale for the personalized virtual coach 'Denk je zèlf!'*

The virtual coach 'Denk je zèlf!' (Dutch for 'Develop a wise mind and counsel yourself') is an interactive and self-learning persuasive system that coaches in attitude and/or long-term behavior change by providing real-time personalized support [7]. Based on the context (e.g., location), emotional state of the user, and natural language processing, the virtual coach application enables tailoring of the real-time feedback to the individual user. The coach application communicates with the user over a chat timeline and provides personal feedback based on persuasive coaching strategies, originated from validation and dialectical strategies of DBT and persuasive features of the Persuasive System Design model [19].

The open-source natural language parser, Alpino [23] is self-learning, able to recognize and understand sentences that contain emotional words or phrases. The emotion-labeled words will be divided over the basic 6 emotions [24].

The virtual coach system consists of four educational modules on intake and commitment, mindfulness, emotion regulation and stress tolerance. The user is guided to set a personal profile, and set preferred moments for daily reminders via a text message. Users are encouraged to keep track of the moments they are experiencing cravings or giving in to a binge, and what emotions they are experiencing at those moments. The coaching system will be designed for an Android smartphone platform.

III. APPROACH AND METHODS

A. *Research protocol*

For the purpose of this two week field research study, we design a mockup version of the virtual coach in order to evaluate validating and dialectical strategies with the target users.

This research study aims at answering the following research questions: "Which coaching strategies do users with a specific type of emotional eating behavior benefit most from while consulting their personalized virtual coach?; "Which coaching strategies are optimal for which emotions?" and "Which coaching approach do users prefer in which context, e.g., time of the day, before/after a craving?"

B. *Target group and participants*

We focus on emotional eaters (N=30), age 18-45 years, both women and men. Participants will be recruited via a Dutch franchise organization of dietitian nutritionists, who are specialized in treating emotional eating behaviors.

C. *Procedures*

First, a pilot study will be conducted with 2-3 participants to validate the study protocol and to test the virtual coach prototype application. Participants are ensured in the anonymity of their participation. Data collected during this research study will be analyzed for research purposes only. The data will be stored anonymously and safely on a local server, using the encrypted secure network connection (https) only the main researchers have access to.

D. *Questionnaires and measures*

Participants will be asked to fill out the demographics data ((nick) name, age, gender, weight, length, place of residence) and three questionnaires on eating behaviour, personality and quality of life:

- the Dutch Eating Behavior Questionnaire (DEBQ) [25]
- Five Factor Personality Inventory (FFPI) [26]
- Positive and Negative Affect Scales (PANAS) [27].

E. *Ecological Momentary Assessments*

During the whole time span of the research study, namely 14 days, participants will be reminded by a text message (SMS) on a daily basis to visit the mobile application and answer two questions: "Did you have cravings / have a binge today?' and "How do you feel at the moment?" The answers to these questions will allow us to find out what the optimal coaching moments might be.

F. Behavioral Chain Analysis

In the behavioral chain analysis (diary), participants will be asked to fill in what happened before they gave in to a binge, and which emotions/feelings they were experiencing.

G. Dialectical Dialogues

Participants will be presented with short dialogues (validating or dialectical) and they will be asked to select the ones they prefer, according to their (current) emotions. To trigger a certain emotion (e.g., the affect that fits best with the chosen coaching strategy), a set of pictures will be shown to the user that evoke respectively sadness, anger, fear, and disgust [28]. Then, a user will be asked to select one or more emotions they are experiencing at the moment they saw the picture(s). After that, dialogues that fit with the dominating emotion, will be presented to the user. The questions in the dialogues are labeled with one of the 4 basic negative emotions (anger, sadness, fear, disgust) and the answers are presented in accordance with one of the validating and/or dialectical (dissonant) strategies.

H. Post-interviews

Participants will be invited to participate in a semi-structured post-interview to discuss their experiences with the presented dialogues and their motivation for picking specific answers. Post-interviews will be audio recorded with permission of participants.

IV. DISCUSSION, CONCLUSION AND FUTURE WORK

This article presents the research study design to evaluate and validate the coaching strategies that emotional eaters prefer in a personalized virtual coach, based on their emotional state and context. The potential contribution of this research is divided into two fields.

First, a multidisciplinary holistic user centered design approach provides valuable insights into preferred persuasive coaching strategies of the target users for developing a meaningful and persuasive virtual coaching system for self-managing emotions. One of the main goals of the personalized virtual coach for emotional eaters is to raise awareness of their own emotions and to enhance a positive change of attitude towards accepting the negative emotions they experience daily. Ultimately, enhancing emotion regulation is expected to lead to a better weight management of emotional eaters.

Second, it will gain more insights into applying the behavior change techniques from health psychology and persuasive technology for virtual coaching application. Next step is to focus on developing virtual coaching systems that actually coach the user by analyzing real-time behavior and providing real-time personalized timely feedback based on user profile, events and context. Coaching based applications can provide self-management support for patients at times when a therapist is not available, or in the case that the user needs personalized anonymous support that is always within reach, which is the case for emotional eaters who often experience feelings of shame.

To conclude, this study makes two main contributions: (1) gaining more insights into: a) what coaching strategies emotional eaters prefer; b) what triggers emotional eaters to (over)eat and at what times of the day the craving usually occur; c) emotional states at the moments of experiencing food cravings and giving into binges, and (2) we aim at developing a general personalized eCoaching framework, enabling the optimal translation of successful behavior change techniques, such as dialectical strategies, into persuasive coaching strategies implementable for broader target groups.

REFERENCES

[1] World Health Organisation WHO | Obesity and Overweight. January (2015) Available at: http://www.who.int/mediacentre/factsheets/fs311/en/ [retrieved: January 31, 2017]

[2] T. Van Strien, C.P. Herman, and M.W. Verheijden, "Eating style, overeating and weight gain. A prospective 2-year follow-up study in a representative Dutch sample," Appetite, 12;59(3), pp. 782-789, 2012.

[3] J. Harvey-Berino et al. "Internet delivered behavioral obesity treatment," Prev Med, 08;51(2), pp. 123-128, 2010.

[4] M.J. Hutchesson et al. "eHealth interventions for the prevention and treatment of overweight and obesity in adults: a systematic review with meta-analysis," Obesity Reviews, 16, 5, pp. 376-392, 2015.

[5] G.M. Manzoni, F. Pagnini, S. Corti, E. Molinari, and G. Castelnuovo, "Internet-based behavioral interventions for obesity: An updated systematic review," Clinical Practice and Epidemiology in Mental Health, 7, pp. 19-28, 2011.

[6] M.M. Linehan, "Skills training manual for treating borderline personality disorder," New York, NY US: Guilford Press, 1993.

[7] A. Dol, O.A. Kulyk, H. Velthuijsen, J.E.W.C. van Gemert-Pijnen, and T. van Strien, "Developing a Personalised Virtual Coach 'Denk je zèlf!' for Emotional Eaters through the Design of Emotion-enriched Personas," International Journal On Advances in Life Sciences, vol 8, 3&4, 2016. *accepted for publication*

[8] H. Geissler, M. Hasenbein, S. Kanatouri, and R. Wegener, "E-Coaching: Conceptual and Empirical Findings of a Virtual Coaching Programme International," Journal of Evidence Based Coaching and Mentoring. Vol 12 (2), pp. 165 – 187, 2014.

[9] S. Kelders, O.A. Kulyk, L. van Gemert-Pijnen, and H. Oinas-Kukkonen, "Selecting effective persuasive strategies in behavior change support systems," In: Third International Workshop on Behavior Change Support Systems (BCSS 2015), In conjunction with PERSUASIVE 2015, June 3-5, Chicago, USA, 1 – 6, 2015.

[10] N. Beerlage-de Jong, A. Eikelenboom-Boskamp, A. Voss, R. Sanderman, and J. van Gemert-Pijnen, "Combining User-Centered Design with the Persuasive Systems Design Model; The Development Process of a Web-Based Registration and Monitoring System for Healthcare-Associated Infections in Nursing Homes," Int. Journal on Advances in Life Sciences, 6, pp. 262-271, 2014.

[11] O.A. Kulyk, C. den Daas, S. David, and L. van Gemert-Pijnen, "How Persuasive are Serious Games, Social Media and mHealth Technologies for Vulnerable Young Adults? Design Factors for Health Behavior and Lifestyle Change Support: Sexual Health Case," In: Proc. of the 3d Int. Workshop on Behavior Change Support Systems (BCSS 2015). In conjunction with PERSUASIVE 2015, June 3-5, Chicago, USA, 2015.

[12] J. Weizenbaum, "ELIZA--A Computer Program For the Study of Natural Language Communication Between Man and Machine," Communications of the ACM. Volume 9, Number 1, pp. 36-35, 1966.

[13] T.W. Bickmore, D. Schulman, and C. Sidner, "Automated interventions for multiple health behaviors using conversational agents," Patient education and counseling 92(2), pp. 142-148, 2013.

[14] O.A. Kulyk, H.J.A. op den Akker, R. Klaassen, and J.E.W.C. van Gemert-Pijnen, "Personalized Virtual Coaching for Lifestyle Support: Principles for Design and Evaluation," International Journal on Advances in Life Sciences, 6;3&4, pp. 300-309, 2014.

[15] A. Van Wissen, C. Vinkers, and A. van Halteren, "Developing a Virtual Coach for Chronic Patients: A User Study on the Impact of Similarity, Familiarity and Realism," Proceedings 11th International Conference, PERSUASIVE 2016 Salzburg, Austria, April 5–7, 2016.

[16] E. Hudlicka, "Virtual training and coaching of health behavior: Example from mindfulness meditation training," Patient Education and Counseling. 92, pp. 160–166, 2013.

[17] R.J. Beun, et al., "Improving Adherence in Automated e-Coaching. A case from Insomnia Therapy," Proceedings 11th International Conference, PERSUASIVE 2016 Salzburg, Austria, April 5–7, 2016.

[18] A. Lentferink et al., "Self-tracking and Persuasive eCoaching in Healthy Lifestyle Interventions: Work-in-progress Scoping Review of Key Components," Proceedings of the Fourth Int. Workshop on Behavior Change Support Systems (BCSS16), co-located with the 11th Int. Conference PERSUASIVE 2016, Salzburg, Austria, April 5, pp. 15-35, 2016.

[19] H. Oinas-Kukkonen, and M. Harjumaa, "Persuasive Systems Design: Key Issues, Process Model, and System Features. Communications of the Association for Information Systems," 24(1), 28, pp. 485-500, 2009.

[20] T.R. Lynch, A.L. Chapman, M.Z. Rosenthal, J.R. Kuo, and M.M. Linehan, "Mechanisms of change in dialectical behavior therapy: Theoretical and empirical observations," J Clin Psychol. 04;62(4), pp. 459-480, 2006.

[21] S. Shiffman, A.A., Stone, and M.R. Hufford, "Ecological momentary assessment," Annu Rev Clin Psychol, 4:1–3.6, 2008.

[22] I. Myin-Germeys et al., "Experience sampling research in psychopathology: opening the black box of daily life," Psychological Medicine, 39, pp. 1533-1547, 2009.

[23] "PIONIER Project," University of Groningen, The Netherlands. Available at: http://www.let.rug.nl/vannoord/alp/Alpino/ [retrieved: January 31, 2017].

[24] P. Ekman, "An argument for basic emotions," Cognition and Emotion, 6(3-4), pp. 169-200, 1992.

[25] T. Van Strien, J.E.R. Frijters, G.P.A. Bergers, and P.B. Defares, "The Dutch Eating Behavior Questionnaire (DEBQ) for Assessment of Restrained, Emotional, and External Eating Behavior," International Journal of Eating Disorders, 5(2), pp. 295-315, 1986.

[26] A.A. Hendriks, K.B. Hofstee, and B. de Raad, "The Five-Factor Personality Inventory (FFPI)," Personality and Individual Differences, 27 (2): pp. 307-325, 1999.

[27] D. Watson, L.A. Clark, and A. Tellegen, "Development and validation of brief measures of positive and negative affect: The PANAS scales," Journal of Personality and Social Psychology, Vol 54(6), pp. 1063-1070, 1988.

[28] K.S. Quigley, K.A. Lindquist, and L.F. Barrett, "Inducing and measuring emotion: Tips, tricks, and secrets," Chapter in H. T. Reis and C. M. Judd (Eds.) Handbook of Research Methods in Social and Personality Psychology, New York, Cambridge University Press, pp. 220-250, 2013.

Toward a Generic Personalized Virtual Coach for Self-management: a Proposal for an Architecture

Johan Blok, Aranka Dol, Talko Dijkhuis

School for Communication, Media & IT
Hanze University of Applied Sciences
The Netherlands, Groningen
e-mail: j.blok@pl.hanze.nl, a.dol@pl.hanze.nl, t.b.dijkhuis@pl.hanze.nl

Abstract—**The shift toward prevention and self-management in health hinges upon a massive realization of behavior change, which involves the use of virtual coaches. Increasing availability of data from mobile devices and techniques for real time analysis provide new opportunities for personalizing virtual coaches. We propose an architecture that takes advantage of those developments. We identify the required knowledge and methods to develop a flexible ecosystem for rapid prototyping of personalized virtual coaches.**

Keywords–virtual coach; personalization; coaching strategy; data science; model based software development.

I. Introduction

Self-management is widely seen as a viable contribution to sustainable health-care as it allows to prevent health problems. Virtual coaches are important means to support self-management in order to maintain and increase health and well-being.

The wide availability of wearables and new techniques of data science to deal with big data provide many new opportunities for the persuasive design of virtual coaches [1]. The variety in health related data is increasing. While motion trackers have become common, other wearables are becoming available, such as stress sensors. Yet, many applications are still at a stage in which the data are simply presented as graphs. Mostly, data are merely used as input to a monitoring tool rather than as input for coaching strategies, let alone that the variety of data is combined in order to personalize the coaching strategy. While van Gemert-Pijnen et. al. and Siewiorek et. al. identify these opportunities [1][2], we propose an architecture for integrating big data in virtual coaches. The prototype described by Blaauw et al. combines the execution of ecological momentary assessments (EMA) with the collection of data from wearables [3]. In our proposal, we embed those two elements within a coaching strategy. Additionally, we enrich the notion of EMA by allowing for automatic execuction and personalization of EMAs. Albaina et al. illustrate the design process of a persuasive virtual coach for the very specific purpose of motivating elderly people to walk by using a pedometer and goalsetting [4].

This paper proposes an architecture for a generic personalized virtual coach. Section 1 introduces the main fields of knowledge that are required for developing a personalized virtual coach. Subsequently, we propose an architecture integrating knowledge of these disciplines. Section 3 elaborates on the required computer science techniques. The next section shortly describes the kind of process management that is required in order to take the concept of virtual coaching a step further. A discussion of future work finishes this paper.

II. Methodology

We define a virtual coach as a computational system that assists the user, either as stand in or merely as supplementary to a human coach, to support behavior that is desired by the user to improve health or well-being. This definition is wider than the definition of eCoaching by Lentferink [5], even if it is narrowed down with the adjective 'personalized' as defined below. Contrary to Lentferink's definition, which limits eCoaching to goalsetting, our definition is agnostic to the nature of the coaching strategy. As a result, we are able to distinguish between the generic nature of the virtual coach and a wide range of coaching strategies. Although the architecture as such does not exclude a wider application, it is developed within the context of eHealth. The intended applications range from the treatment of emotional eaters [6][7] to support for maintaining the social network by elderly people. As such, the virtual coach implements persuasive technology that attempts to influence the behavior of its users.

According to the behavior model of Fogg, behavior is most likely to be influenced when the motivation is high and the triggers to act are easy to execute [8]. Given the intrinsic difficulties of achieving enduring behavior change, our definition deliberately limits the virtual coach to changes that are desired by the user. According to Fogg's behavior model, triggers become easy to execute when they are adapted to the context of the user. Fogg's characterization of triggers straightforwardly translates into a crucial requirement for the virtual coach, namely that it must personalize its feedback to the user as well as the coaching strategies on which this feedback relies. A particular coaching strategy, or a different focus within a coaching strategy, might be more effective, depending on the dispositions, abilities, preferences, and lifestyle of the user.

The architecture proposed in the following section is based on the integration of three disciplines. The first one belongs to the therapeutical application of behavioural sciences, sociology or psychology, or a mix of them, depending on the nature of the intervention for which one implements the virtual coach. They provide an evidence based coaching strategy. For example, if an instantiation of the virtual coach needs to support the development of learning abilities of children, the coaching strategy will stem from theories about metacognition and the manner in which capabilities can be improved. In the case of emotional eaters, which comprises our first application of the virtual coach, the coaching strategy stems from existing face-to-face therapies [9].

The second discipline is that of data science. The wide spread use of mobile devices and recent developments in the field of big data and predictive analytics allows to achieve the required personalization by gathering, analyzing, and interpreting personal data. Analysis, both on the level of personal data (n=1) and of aggregated data, provides information to guide the coaching strategy in adapting its triggers to the specific circumstances of the individual user. Most importantly, streaming data, predictive analysis, and realtime data processing provide novel opportunities to achieve an appropriate timing of the triggers, which is key in the success of persuasive designs [8].

Finally, the third discipline comprises a subfield of software engineering called model driven software development (MDSD). The application of domain models, especially in the form of domain specific languages (DSL), entails the development of abstractions that capture the crucial elements of a domain, in our case coaching strategies [10]. For the purpose of the virtual coach, a DSL must be designed that allows for an straightforward implementation of coaching strategies. Such an implementation consists of a translation of (parts of) a coaching strategy into rules suitable for a computational system. The virtual coach executes these rules to provide feedback to the user depending on the outcomes of the analysis of data. Since there are many ways in which even a single coaching strategy can be implemented, considerable exploration and research must be carried out into the most effective ways of implementing coaching strategies. The application of MDSD ensures that the rules of a coaching strategy can be changed easily.

III. Architecture

We propose to combine a health data platform (HDP) with a platform for the implementation of coaching strategies (CSP) in order to establish an ecosystem for personalized virtual coaches (see Fig. 1).

A. Health Data Platform

One of the most important aspects of the required personalization of the triggers of a coaching strategy consists in the trigger's timing [8]. The real time analysis of personal data streams as well as the real time self-evaluation of the virtual coach itself provides new opportunities to establish an appropriate timing of the triggers.

The HDP employs techniques of data science to gather, store, query, analyze, classify, and predict health and lifestyle related data. We distinguish between direct and indirect data streams. The former stem from mobile devices, while the latter consist of data that is generated by the coaching strategy itself. The indirect data streams allow the virtual coach to consult the coaching history and adapt itself accordingly, for example to avoid coaching moves that have failed in the past.

B. Coaching Strategy Platform

The other part (CSP) implements the coaching strategy and consists of the following parts:

- user interaction;
- data processing unit to combine data generated by user interaction with the output of the health data platform;
- rule engine to provide the user with feedback by personalizing scheme's.

First of all, feedback may consist of advice to take educational modules to improve skills as required by the coaching strategy. It may also consist of the suggestion to set a certain goal. When accepted by the user, progress will be measured and visualized in such a way that it motivates the user to achieve this goal. A third important type of feedback consists of an EMA.

EMAs are often used in eHealth related research. They provide a convenient way to obtain structured real time information about the behavior and experiences of the user [11][12]. EMAs are used for research purposes as well as treatment. We propose a way of using EMAs within the virtual coach with two novel aspects:

- automatic execution;
- personalization.

Depending on the specific state of the individual, the coaching strategy might involve an EMA at certain stages of the coaching. Schematized EMAs can be used to personalize ecological momentary interventions, or EMIs. Coaching rules enable the automatic execution of EMIs that are tailored to individual situations.

Let us take a simple example to illustrate how this global architecture supports the implementation of a virtual coach. Suppose a wearable is used for tracking the movements of users. Let us take as a very simple coaching strategy that the virtual coach must be able to support existing good habits. Pattern detection and predictive analytics learns for example that a particular user A is usually running on each Friday morning. Let us assume that the user has entered his holidays via the user interface or that the virtual coach has access to this information by other means. and that he spends his Saturdays such that it is no option to improve movements. This extremely simplified coaching strategy can be implemented in the rule engine as follows:

```
if running is predicted within 5 minutes,
remind the user unless he has a holiday
```

The user will receive notice in accordance with her or his preferences (SMS, whatsapp, etc.). An improved version might take into account the weather and only remind the user when it has stopped raining.

IV. Computer Science Techniques

This section identifies the required techniques of the most innovative aspects of the proposed architecture.

A. Data Science

Widely available mobile devices provide diverse data streams that are useful for understanding, and possibly predicting, the behavior of the user. The nature of the virtual coach requires the analysis of these data streams to be automated. Patterns, features, and anomalies need to be discovered automatically (knowledge discovery). Knowledge discovery requires:

- the automation of the data analysis process;
- the handling of Big Data in combination with Analytics.

The automation of data analysis involves the preparation, transformation, and modeling of data in order to become in the position to discover 'knowledge', that is, useful information [13]. The diversity, realtime, and streaming nature of the data impose Big Data aspects [14] on the automation of data analysis [15]. Accordingly, the health data platform requires the latest techniques and frameworks to implement the storage and real time analysis of data streams, such as Apache Storm.

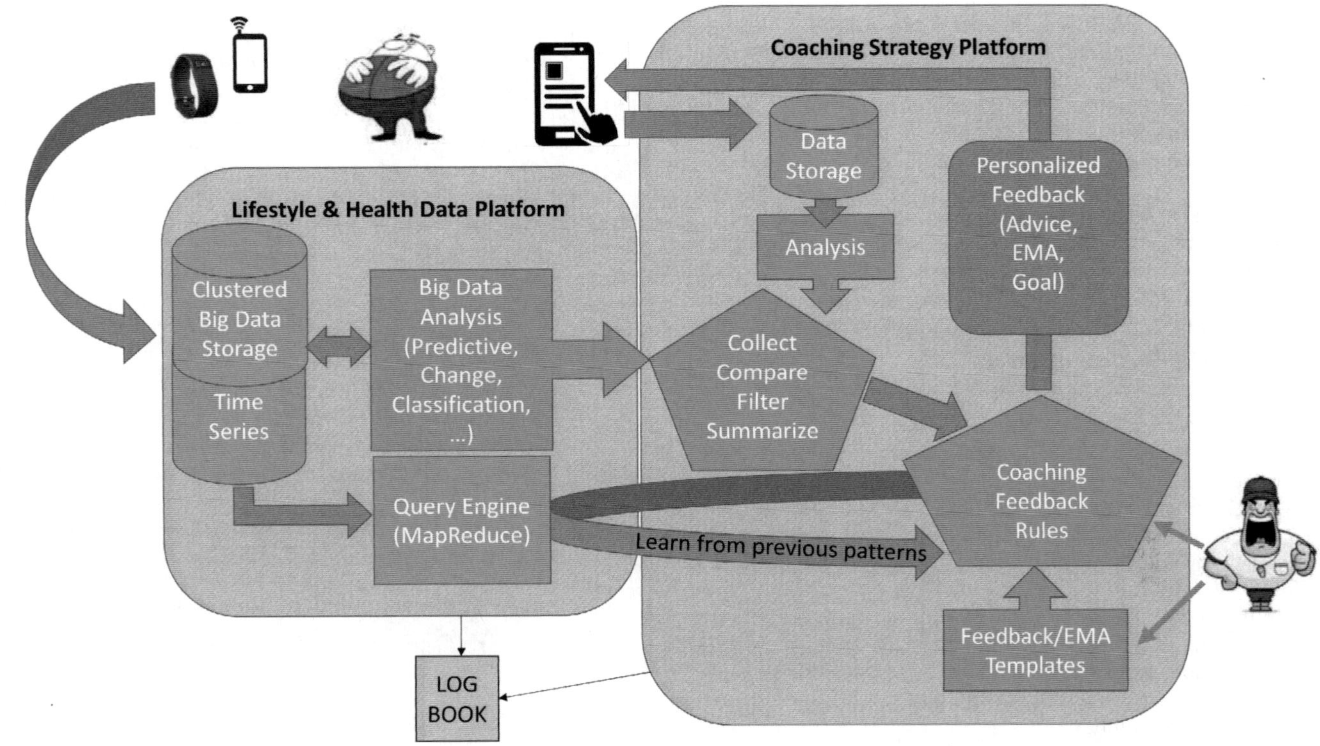

Figure 1. Architecture Personalized Virtual Coach

B. Model Driven Software Engineering

The implementation of a coaching strategy requires at least some techniques that are developed under the label of artificial intelligence (AI). A common distinction between two approaches in AI is that between rule based expert systems and data based machine learning systems. Currently the latter is quite popular, because of success in applying this approach to previously unsuccessful cases, such as machine translation and computer vision. To a large extent, this success is the result of the availability of a lot of data of sufficient quality, such as labeled images for computer vision.

Unfortunately, in the case of coaching strategies, this data is hardly available (or not in the language of the intended users) and is not likely to become available in the foreseeable future. Moreover, experts on coaching strategies aim at improvements, which requires a conceptualization of psychological and/or social processes. In the case of machine learning, however, correctly labeled data is used to produce rules for computation indirectly. These automatically produced rules not only remain implicit, but also remain almost incomprehensible for human cognition. Thus, a machine learning solution to the coaching strategy as a whole tends to obfuscate an understanding of these processes, which is precisely what domain experts aim to unravel. Although machine learning is invaluable to analyze the data streams, and can be used in several other parts, such as natural language processing, we do not regard it as the first choice for implementing the coaching strategy as a whole. Nevertheless, once sufficient virtual coaches, using sufficiently rich data sets, have operated for some time, their data might be useful to train machine learning modules to improve the virtual coaches. Currently, however, we are not yet at this stage, but have to start by implementing coaching strategies as understood by domain experts.

The field of MDSD provides convenient means, such as the concept of a DSL, to implement a rule based system. Such a DSL based system allows to formulate the coaching strategy in terms that are close to the coaching domain. In the case of dialectical behavior therapy for emotional eaters, a simple example of such a rule could be [9]:

```
if Anger or Fear dominates
follow Dialectical Coaching Strategy
```

Obviously, such a rule presupposes several software modules, such as modules that allow to determine the dominating emotions, for example by means of EMAs, a stress sensor, or the analysis of expressions in natural language. The DSL provides an abstraction level that does not require knowledge about the way these software modules and the analysis of data are realized.

V. Development process

Nowadays, agile approaches dominate the development of software. These approaches ensure a crucial role for all stakeholders at each stage of the development. An agile approach, especially in the form of scrum, is particularly suitable for a development process that requires the integration of quite diverse disciplines. By means of an, preferably daily, participation in the scrum process, the expert on coaching strategies is immediately confronted with the the (im)possibilities of engineering techniques Conversely, the engineers get immediate feedback on their interpretation of the coaching strategy. In addition, the agile approach prepares the virtual coach for adoption by users and the healthcare sector [16].

VI. Conclusion

We have identified how knowledge of diverse disciplines should be integrated, and proposed an architecture for a generic personalized virtual coach that utilizes recent developments in wearables and data science. Focusing on the implementation of the coaching strategy required by the treatment of emotional eaters, several software modules for both the HDP and CSP are currently under development. Future work will consist of the integration of the software modules to show the viability of our architecture, including the design of an appropriate DSL for implementing coaching strategies. Planned expansions to other coaching cases will warrant and demonstrate the generic nature of our solution.

References

[1] L. V. Gemert-Pijnen, F. Sieverink, L. Siemons, and A. Braakman-Jansen, "Big data for personalized and persuasive coaching via self-monitoring technology," 2016.

[2] D. Siewiorek, A. Smailagic, and A. Dey, "Architecture and applications of virtual coaches," *Proceedings of the IEEE*, vol. 100, no. 8, pp. 2472–2488, 2012.

[3] F. J. Blaauw *et al.*, "Let's get physiqual - an intuitive and generic method to combine sensor technology with ecological momentary assessments," *Journal of Biomedical Informatics*, vol. 63, pp. 141–149, 2016.

[4] I. M. Albaina, T. Visser, C. v. d. Mast, and M. H. Vastenburg, "Flowie: A persuasive virtual coach to motivate elderly individuals to walk," pp. 1–7, 2009.

[5] A. Lentferink *et al.*, "Self-tracking and persuasive eCoaching in healthy lifestyle interventions: Work-in-progress scoping review of key components," pp. 15–35, Apr. 5, 2016.

[6] A. Dol, O. Kulyk, H. Velthuijsen, L. V. Gemert-Pijnen, and T. V. Strien, "Denk je zèlf! developing a personalised virtual coach for emotional eaters using personas," pp. 42–47, 2016.

[7] ——, "Developing a personalised virtual coach 'denk je zèlf!' for emotional eaters through the design of emotion-enriched personas," *International Journal On Advances in Life Sciences*, vol. 8, no. 3 & 4, 2016, accepted.

[8] B. Fogg, "A behavior model for persuasive design," Persuasive '09, 40:1–40:7, 2009.

[9] A. Dol, "Translating dialectic dialogues into persuasive coaching strategies for personalized virtual coach: A research protocol for emotional eaters case study," forthcoming.

[10] M. Voelter, *DSL engineering: Designing, implementing and using domain-specific languages*. CreateSpace Independent Publishing Platform, 2013, 558 pp.

[11] S. Shiffman, A. A. Stone, and M. R. Hufford, "Ecological momentary assessment," *Annual Review of Clinical Psychology*, vol. 4, pp. 1–32, 2008.

[12] I. Myin-Germeys, M. Oorschot, D. Collip, J. Lataster, P. Delespaul, and J. v. Os, "Experience sampling research in psychopathology: Opening the black box of daily life," *Psychological Medicine*, vol. 39, no. 9, pp. 1533–47, 2009.

[13] J. Shyr and D. Spisic, "Automated data analysis," *Wiley Interdisciplinary Reviews: Computational Statistics*, vol. 6, no. 5, pp. 359–366, 2014.

[14] A. Gandomi and M. Haider, "Beyond the hype: Big data concepts, methods, and analytics," *International Journal of Information Management*, vol. 35, no. 2, pp. 137–144, 2015.

[15] A. Labrinidis and H. V. Jagadish, *Proceedings of the VLDB Endowment*, vol. 5, no. 12, pp. 2032–2033, 2012.

[16] J. Baljé, A. Carter, and H. Velthuijsen, "Agile development as a change management approach in healthcare innovation projects," 2015.

Prediction of Running Injuries from Training Load: a Machine Learning Approach.

Talko Dijkhuis[1,3], Ruby Otter[2,4], Hugo Velthuijsen[1] and Koen Lemmink[4]
[1]Hanze University of Applied Sciences,Institute for Communication, Media & IT, Groningen, The Netherlands
[2]Hanze University of Applied Sciences,Institute of Sport Studies, Groningen, The Netherlands
[3]University of Groningen, Johann Bernoulli Institute for Mathematics and Computer Science, Groningen, The Netherlands
[4]University of Groningen, Center for Human Movement Sciences, University Medical Center Groningen, Groningen, The Netherlands
[1]{t.b.dijkhuis, t.a.otter, h.velthuijsen}@pl.hanze.nl
[4]{k.a.p.m.lemmink}@rug.nl

***Abstract*—The prediction of the running injuries based on self-reported training data on load is difficult. At present, coaches and researchers have no validated system to predict if a runner has an increased risk of injuries. We aim to develop an algorithm to predict the increase of the risk of a runner to sustain an injury. As a first step Self-reported data on training parameters and injuries from high-level runners (duration=37 weeks, n=23, male=16, female=7) were used to identify the most predictive variables for injuries, and train a machine learning tree algorithm to predict an injury. The model was validated by splitting the data in training and a test set. The 10 most important variables were identified from 85 possible variables using the Random Forest algorithm. To predict at an earliest stage, so the runner or the coach is able to intervene, the variables were classified by time to build tree algorithms up to 7 weeks before the occurrence of an injury. By building machine learning algorithms using existing self-reported training data can enable prospective identification of high-level runners who are likely to develop an injury. Only the established prediction model needs to be verified as correct.**

Keywords–Human Performance; Machine Learning; Predictive Analysis; Load; Injuries; Monitoring; Endurance Athletes

I. INTRODUCTION

In the context of the project 'Groningen Monitoring Athletic Performance Study'(MAPS) [1] two years of data was gathered about load, and tests on the performance of competitive athletes. The investigation into the factors that influence performance and injury risk of athletes lead to more insight. The effect of a change in load on injuries is difficult to predict. The knowledge on the relationship between load and the effect on injuries might be improved by matching self-reported training data and injury data using machine learning techniques. At present, Human Movement Researchers of the project MAPS have no published findings on the relation between a change in load and injuries on endurance athletes. To identify the predictors, a choice has to be made on which information is to be used for developing an algorithm for the prediction of injuries. There are two kinds of data available, the daily self-reported training log data and the injury data of high-level competitive runners. An injury is defined as any musculoskeletal problem of the lower extremity or back that lead to an inability to execute training or competition as planned for at least one week [2]. The training log data contains information about the training duration and the training intensity. The intensity times the duration is the workload of a training. The terms acute and chronic workload are used to descibe the intensity of the immediate window of training. Acute workload is the average workload of the last seven days, chronic workload is the average workload of the last 28 days. [3]. A study to sustain an injury risk of rugby league players was conducted with the acute:chronic ratio and concluded that there was a relation between High Chronical workload, spikes in the acute workload and the increased risk on injuries [4]. To find a relation between the workload and injuries for high-level runners, the data of the training data log is to be converted to predictors, which are based on the acute:chronic ratio, derived ratio's and the workload. Machine Learning is an appropriate manner for examination of all these predictors [5] because Machine Learning Techniques can discover complex high dimensional interactions between predictors and predict the label of injury/no injury.

II. METHODS

A. Study design

The study design stems out of the MAPS project. A prospective cohort study was used in which 23 high-level competitive runners (16 male, 7 female) were followed for 14 months. During this period they reported data on training parameters and injuries(Table I).

TABLE I. BASELINE CHARACTERISTICS OF RUNNERS

Characteristic	*Male*	*Female*	*Total*
Number	16	7	23
Age(years;mean +- SD)	22.5 +- 6.3	21.4 +- 4.4	22.2 +- 5.7
Height(cm;mean +- SD)	185 +- 5	172 +- 7	181 +- 8
Body weight(kg;mean +- SD)	68.6 +- 6.0	58.3 +- 4.0	65.4 +- 7.2
Perc. body fat(perc.;mean +- SD)	8.5 +- 2.3	17.6 +- 4.2	11.3+- 5.2
VO2max(ml/min/kg)	66.7 +- 5.9	62.7 +- 7.4	65.5+-6.5

B. Dataset description

1) Overall dataset: The runners kept a daily training log, in which, information about the training duration, the training intensity and the sustained injuries was reported. The dataset contains 208 training patterns of 7 weeks, 52 patterns with injuries and 156 patterns without any injury. To find a relation between the workload and injuries for high-level runners, the data of the training data log is converted to features, which are based on the acute:chronic ratio, monotony, strain and the workload. For a sliding window of 7 weeks before the injury, every single week the same ratio's and workload were calculated. The final step was to determine the percentage of change between the features of the identified weeks and add this to the dataset

2) Injuries: There were 22 runners, out of 23, injured during the observed period of 14 months. Only the complete patterns were selected for the dataset. These are the athletes who had no missing data in their daily training log in combination with the reported injuries. And only the injuries, which had a sliding window of 7 weeks of self-report data before the injury were used.

3) No injuries: The period before the injury has to be compared with the period in which the workload doesn't lead to injuries. Therefore the 7 week period without resulting into an injury, before the 7 weeks sliding window of the injury date, was taken for every individual athlete.

C. Model development

For the identification of the most important features Random Forest of Sklearn was used. Random Forest builds trees on subsets of the data, bagging, and there for applicable for relative small datasets. [6] Next, the relative importance of the features can be identified. To intervene as early as possible, single trees will be assembled based on the moment in time using the identified features.

D. Feature selection

The features within and between the sliding windows are correlated with each other. The following features are identified as the most important, sorted by decreasing importance.

1) average workload week 2
2) sum workload week 2
3) percentage change monotony between week 1 and 2
4) acute:chronic ratio 7 over 42 week 7
5) acute:chronic ratio 7 over 28 week 7
6) percentage change strain between week 1 and 2
7) percentage change workload between week 2 and 3
8) acute:chronic ratio 7 over 42 week 2
9) percentage change strain between week 2 and 3
10) strain 2

The Random Forest is also used to predict on the dataset and had predicted 67 percent of the injuries correct.

E. Predictive modelling building

The 10 features are used to build individual trees to be able to identify as early as possible the increase of risk. The dataset was divided random in a training and a test set in respectively 75% and 25%. Building a tree on the features 'acute:chronic ratio 7 over 42 week 7' and 'acute:chronic ratio 7 over 28' resulted in an accuracy of 75 %. With these ratios the prediction of the injury was 75% correct. Other trees are not built yet.

III. Results

We selected 10 predictors to predict the occurrence of an injury in the future over an sliding window of 7 weeks. The predictive modelling is promising but it is also a bit suspicious that an tree is more accurate using 20% of the features in comparison with the Random Forest.

IV. Conclusion

The predictive modelling using two steps in the process seems to be promising. But the accuracy of the tree on 20% of the feature set is very high, future work is to investigate the rationale behind the results. The established prediction model needs to be verified as being correct. When a robust and accurate prediction model is realized, the model will be added to an app. The app will get information about the workload per training session and, using the model, predicts the probability on sustaining an injury.

Acknowledgment

The authors would like to thank Henk van der Worp for identifying the injuries in the data of the runners and M. Aiello for suggesting on improving the paper.

References

[1] S. Raak-Pro, "Belasting en belastbaarheid van topsporters," 2011. [Online]. Available: http://www.sia-projecten.nl/projectenbank/project/belasting-en-belastbaarheid-van-topsporters

[2] S. W. Bredeweg, S. Zijlstra, and I. Buist, "The GRONORUN 2 study: effectiveness of a preconditioning program on preventing running related injuries in novice runners. The design of a randomized controlled trial," BMC Musculoskelet Disord, vol. 11, 2010, p. 196. [Online]. Available: http://www.ncbi.nlm.nih.gov/pmc/articles/PMC2936887/pdf/1471-2474-11-196.pdf

[3] T. J. Gabbett, B. T. Hulin, P. Blanch, and R. Whiteley, "High training workloads alone do not cause sports injuries: how you get there is the real issue," British Journal of Sports Medicine, vol. 50, no. 8, 2016, pp. 1–2. [Online]. Available: http://bjsm.bmj.com/lookup/doi/10.1136/bjsports-2015-095567

[4] B. T. Hulin, T. J. Gabbett, D. W. Lawson, P. Caputi, and J. a. Sampson, "The acute:chronic workload ratio predicts injury: high chronic workload may decrease injury risk in elite rugby league players," British Journal of Sports Medicine, vol. 50, no. 4, 2016, pp. 231–236. [Online]. Available: http://www.ncbi.nlm.nih.gov/pubmed/265110065Cnhttp://bjsm.bmj.com/lookup/doi/10.1136/bjsports-2015-094817

[5] M. Kuhn and K. Johnson, Applied predictive modeling. Springer, 2013, vol. 26.

[6] L. Breiman, "Random forests," Machine learning, vol. 45, no. 1, 2001, pp. 5–32.

Improving The Well-Being of Older People by Reducing Their Energy Consumption Through Energy-Aware Systems

Jon Robinson Kofi Appiah Raheem Yousaf

Department of Computing & Technology
Nottingham Trent University
Nottingham, UK
email:jon.robinson@ntu.ac.uk
email:kofi.appiah@ntu.ac.uk
email:raheem.yousaf022012@my.ntu.ac.uk

Abstract-Fuel poverty is becoming a problem amongst the older community in the UK. We propose an assistive technology for reduced energy consumption in aid of active aging. We introduce how the well-being of older people within the community can be improved by reducing anxiety associated with fuel poverty. A recommender and behavioural change system will be introduced, which enables older people to improve their energy footprint through energy-aware systems. Using systems to help reduce fuel poverty will invariably improve their general well-being. Results show how this technology can be accepted and act as an enabler in improving the overall well-being of older people.

Keywords- ***energy-efficiency; energy-awareness; intelligent sensor systems; recommender systems; well-being.***

I. INTRODUCTION

At present, older people within the community face a host of technological and perceptual problems, which can inhibit their interaction and well-being while using information and communications technology (ICT) solutions to improve their lives. This is especially relevant when considering Internet of Things (IoT) based systems to improve their well-being within their home environment, dealing with heating and reducing costs. Fuel poverty is a key societal problem that impacts on large numbers of the older community within the UK [1] as older people tend to fall into the low-income range due to low pensions and rising energy costs. The worry and anxiety concerning their limited financial resources results in 20,000 – 25,000 deaths in the UK [3]. This is due to reducing their dependency on heating their homes as they are not financially secure enough to afford or pay for their monthly house hold heating or cooling bills [4]. The use of technology to help older people reduce costs is constrained by their fear and distrust of technology. To enable older people to accept ICT solutions, the UK government has a number of initiatives in place to help reduce this fear (e.g., digital by default [2]), which aims to provide them with skills to use modern technology. However, figures show that high numbers of individuals are still susceptible to fuel poverty which is not only reducing their well-being but also putting their lives at risk [3].

Compounding the issue of fuel poverty, older members of the community also fear the use of technology which can impact on the successful deployment and adoption of technology to address the issue of fuel poverty. Other issues relating to the acceptance of technology and the security of personal information also factor into the distrust of ICT, which is there to help improve their well-being.

In this work we intend to provide a prototype system to show how unobtrusive and intelligent technology can help reduce the energy consumption of older people and how this can impact on their behaviour by adopting more energy conscious patterns through a real-time responsive recommender system. We will therefore examine whether an *intelligent recommender system can reduce the energy consumption of older people by predicting and modifying their daily patterns and behaviours so that they minimise their energy needs.* The rest of this paper is structured as follows: Section 2 provides an insight into existing systems; Section 3 discusses the design considerations that need to be taken into account when considering energy-awareness as a whole; Section 4 introduces our proposed system and its key components; Section 5 discusses user interaction issues and factors that can deter engagement; Section 6 provides initial results; and finally, in Section 7, we present our conclusion.

II. RELATED WORK

Reducing the energy consumption of individuals has been widely tackled by a variety of commercial and academic systems. An approach taken by some commercial systems is to view each room as a separate entity and give contextual information regarding the state of the room. Some systems require user intervention in the control of the ambient energy consumption. For example, with the *Hive* system [10], to adjust or improve the heating within a room, a separate mobile based application is required which can lead to problems when considering older people. Although additional sensors and smart plugs are available to provide the ability to turn on and off devices, these still need to have some form of user intervention. Voice control or a mobile phone application is required to govern their operation. However, this forces the user to actively monitor, respond and modify temperatures rather than allow the system to proactively shape the energy use of the home. IFTTT (IF This, Then That) [11] is another system which allows IoT integration within the home by providing users with a simple rule-based approach for governing the operation of smart devices. Devices are linked to a web based system which can then control the devices within the home. Systems like this would fail to address the needs and anxiety encountered when older people interact with technology [5] due to mobile technology and applications not being accessible to all older people which also increases their fear of technology. Another similar system is the Honeywell *evoHome* [12] system provides users with contextual information about the room temperature and provides them with a digital thermostat to control ambient temperatures. The *Heat Genius* system [13] is another commercial product which uses motion control to determine whether heating is required in a particular room. This system also provides the ability to learn and adjust room temperatures based on the life style of the users. However, these commercial systems do not address the issue of reducing energy consumption as a whole but instead focus on one specific problem - heating.

Smart energy meters have been used extensively in the home and in some cases targeting low-income and older users. In [15], a number of smart energy systems and the interactions with users are discussed. Here, users used the *Duet* and *Trio* devices in a bid to monitor their real-time energy consumption over a one year period. Smart home and assistive technology can also play an important part in monitoring and adapting the environment to best suit an older person. Kim et al [14] discuss the *U-Health* system for monitoring and supporting the older person within their home. Wireless sensor technologies embedded within appliances in the home allow for the collection and mining of information regarding the habits of the older user, as well as providing decision-making capabilities to allow the system to adapt to the users' needs.

The issue of fuel poverty is a multi-dimensional one which covers more than just the heating of homes. To provide a representative picture of how energy-awareness impacts older people, all aspects of energy consumption needs to be considered. When looking at the problem of fuel poverty, any savings of an older person's energy expenditure must yield in financial benefits. In turn, for a technological solution to fully address the issue of fuel poverty it must holistically look at energy-awareness as a whole and provide ways in which behaviours and usage patterns can be altered to improve energy expenditure through an unobtrusive and proactive system which does not require constant user interaction to govern the control of the environment.

III. DESIGN CONSIDERATIONS

Older people face a number of challenges in today's society which can impact their lives and well-being. Fuel poverty is a major societal challenge which affects significant proportions of the society. With limited financial resources available to older people who are reliant on state pensions, paying for fuel (i.e., electricity and gas) becomes a major concern for their well-being [3]. Within the UK alone, a high number of unnecessary deaths are caused each year due to weather concerns [1]. This can be due to excessively hot weather during the summer months or excessively low temperatures during the winter. In circumstances like this, which require a higher expenditure, it is common for the older person to go without basic heating or cooling due to limited financial resources, which enables them to focus on other costs (e.g., food and rent) [4]. However, this has a detrimental impact on their well-being due to the suffering involved, and possible death, during low or high temperature times during the year. Not only does it impact the older people, but the costs to healthcare for dealing with emergencies relating to the admission and care of older patients at these times, places stress on services which are already over-stretched.

Communities and community care offer care services where carers or social workers visit the older person to ensure that they are coping and not suffering. However, with finite resources, only so much is able to be done. To help in the care of older people, smart technology has been employed quite extensively. For example, in [15], smart devices help monitor the older resident. However, relying on technology can introduce other challenges. More specifically, older people distrust and fear the use of technology [5]. This often leads to technology being ineffective due to the lack of engagement from the older person, especially if they are required to interact and engage with the technology in some way. Davis [5] introduces the Technology Acceptance Model which impacts on the perceived usefulness of technology by the older person which can introduce barriers for adoption. However, older users also encounter problems with a general lack of understanding of ICT which introduces barriers to the user as it impedes them from making strategic decisions when managing fuel costs [9]. This lack of understanding needs to be addressed for the system to be effective and allow seamless interaction with the user. For instance in [14][15] mobile technology is used but can introduce problems with engagement due to the fear of technology whereby the older person has to interact not only with an unknown and unfamiliar device but also some form of application and visual interface. If the interaction between the system and older person is perceived to be too difficult then a lack of engagement usually results [5][9]. Therefore, intelligent devices need to be unobtrusive as well as the system offering a less technological way of interacting with the older person to minimise the impact of the fear of technology.

Behavioural change is another key consideration for such systems as the technology would also have to monitor and determine if there are better behaviour patterns when interacting with energy-reliant devices within the home. For example, making a cup of tea or cooking dinner relies on some form of energy consumption (either electricity or gas) to complete the task. However, energy tariffs can change through the day and can impact on the total expenditure over a month. By modifying the behaviour of the older person to a more cost-conscious time can have benefits in the long term.

Therefore, when designing effective systems for reducing the energy consumption of older people, a number of design considerations have to be taken into account. Namely, the use of unobtrusive technology which is embedded within the environment; simple interaction devices which promote trust; behaviour modification and prediction to reduce or better manage the use of energy reliant devices within the home; and, systems to recommend changes to patterns.

We propose to address the question of whether an intelligent recommender system can reduce the energy consumption of older people by predicting and modifying their daily patterns and behaviours so that they minimise their energy needs to ensure an improvement in their general well-being by reducing anxiety associated with fuel poverty. Anxiety has a clear detrimental effect on the well-being of an older person [8], which will hinder their interactions and acceptance of technology. Improving the quality of life of older users is difficult to gauge as different people assign different values to things [7]. For that reason, the proposed system aims to reduce the fear of poverty and improve the quality of life for older users by providing them with guidance and options on how to change their behaviour so that they are more energy-conscious and aware of what they are doing through a simple house model interface. More detailed feedback can be given to those users who feel comfortable with technology or to carers, or family members who are actively helping the older user. By providing them with a more holistic and wide-ranging understanding of the energy-consumption which takes into account not only their heating but also their activities during the day can help them improve their overall energy consumption. By looking at the whole, more substantial savings can be made which in turn reduce their chances of falling into the fuel poverty trap.

Therefore, in summary we intend to address the issue of the acceptance of technology as well as factors relating to the trust of intelligent systems and improving well-being by reducing fuel poverty. This will be through simple, trusted and unambiguous human-computer interfaces, behaviour and activity analysis; and, recommender systems to help promote more informed energy-aware decisions. This will allow older people to consider all aspects of their energy consumption, rather than focusing on one single aspect (e.g., heating) as their overall energy-footprint will be made out of many different components. For example, heating, cooking, recreation, etc., which complements their potentially sedentary life-style.

IV. PROPOSED SYSTEM

From a conceptual perspective, the system (see figure 1) will process sensor telemetry from devices located within the environment, determine if the behaviour of the person can be changed based on the task activity they are performing, which includes using tariff information gathered from energy suppliers through IoT and then facilitate in providing feedback to the older person.

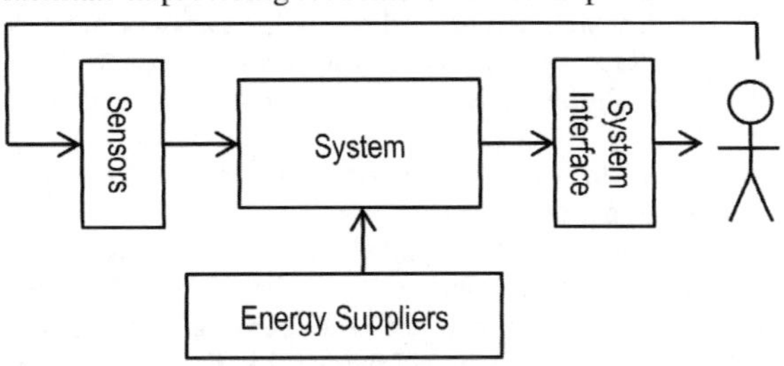

Figure 1. System overview.

The guiding principle in the design of the prototype is to determine if the energy consumption of the user can be reduced by determining the behaviour and patterns they are exhibiting. The system relies upon a number of key components that deal with the interactions with energy suppliers to provide information, which informs the recommender system; activity/task detection and interfacing with the user through an unobtrusive interface. A system prototype has been developed which allows the seamless interaction with an older person, which could reduce their energy consumption. For this prototype iteration the full system was implemented but deployed within an idealised test environment to capture telemetry. The system was also able to save and playback telemetry for user evaluation.

A. *Sensors*

A variety of off-the-shelf sensors have been used to provide contextual information regarding the state of the environment, the location of the older person, and what they are interacting with in the home environment. Sensors that provide telemetry on light, motion, pressure, electricity use, sound, as well as other sensors providing contextual state information of the environment, have been used. Sensors were typically subsumed within the appliances which consumed energy. This was mostly electrical appliances but scope for gas based appliances will be incorporated in a later iteration. Sensors can be added seamlessly to the system by simply plugging them in and informing the software that a particular type of sensor has been added. Different sensors can be dynamically added, along with information which provides contextual information regarding how telemetry can be analysed and interpreted during the behavioural analysis phase. Environmental and power usage sensors were based on Phidgets' technology and were used to record and forward telemetry using an event system that had been developed (details in section IV. B). For the next phase of the work, smart Wi-Fi enabled energy plugs and other Wi-Fi enabled technology, will be used.

B. *Behavioural Analysis*

To aid in the process of determining how much energy is being consumed within the house and whether a better alternative is available, telemetry from sensors are aggregated and processed within the behavioural analysis component of the system. An overview of the activity, behaviour and recommender stages is outlined in figure 2.

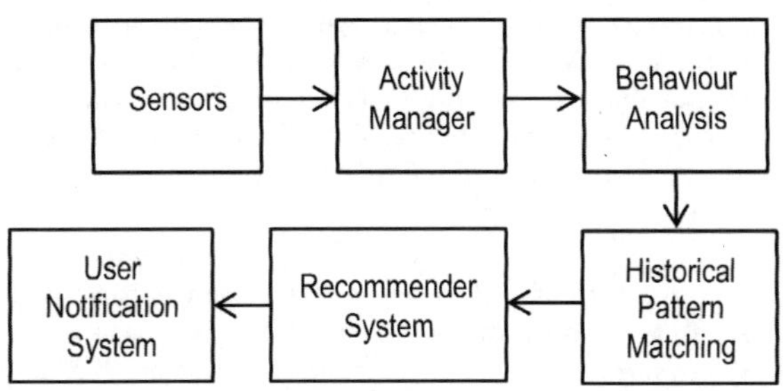

Figure 2. Activity, behavioural and recommender stages.

Activities are used to identify what task the user is performing within the home. For example, making a meal would involve using a variety of sensor equipped devices within the kitchen, each of which will be generating telemetry. In a typical use case, the oven, along with the cooker and supplementary devices (e.g., kettle) could be used. Combined with environmental state information originating from embedded sensors in the home, additionally telemetry regarding light, sound and motion information would be generated and be contextualised within the kitchen environment.

Activities are generated by end-users and allow the association between different telemetry and sensors and a given activity. The approach that has been taken is to offer a simple means for users to define an activity and then to associate sensor combinations to that activity. These activities will then form the basis of the behavioural analysis and pattern-matching functionality whereby the system will be able to determine what type of activity is being performed by the user.

Raw sensor telemetry is tagged with additional meta-information regarding the sensor type, timestamp and location and transmitted around the system within an event message. Events are captured and processed within the behavioural component during the activity identification process. To aid in providing an environmental snapshot of what is happening within the environment, a window of applicability is used which allows events to be received over a short period of time. This populates activity states which are triggered when all associated sensor telemetry has been received during a set period. When this has been satisfied, activities trigger notifications, which are in turn, processed by the analysis component. At this point, a decision can be made on what to do next. The first option is to process an associated action which causes the system to perform some form of real-world action which impacts the end-user. Alternatively, notifications can be used to build up composite events which provide more complex combinations of activities to be associated with each other. This allows the system to provide a richer and more complex activity detection process.

Historical information regarding past activities is used to provide the ability to determine if activities have occurred during similar times in the past. For example, cooking food at specific times; use of kettle to make drinks; and activities that relate to their life-styles. This information is used by the recommender system to determine if user activity patterns can be modified to help improve the energy use within the home based on their occurrences. For instance, when coupled with tariff information, the recommender would be able to inform the user that by changing their pattern to do a particular activity by a few minutes could result in a tangible energy cost saving over a period of time.

C. *Recommender System*

The purpose of the recommender system is to provide end-users with feedback on their energy consumption performance or provide alternative actions. To best inform end-users of any energy cost savings, tariff information is used from the electricity supplier that they are signed up with. By analysing their past behavioural patterns, the recommender system will determine if any energy savings can be made by suggesting to the end-user how best to alter their habits. For instance, as a profile of the user is generated, any patterns or behaviours they exhibit will be identified and any improvements will be suggested. This information can be conveyed in a number of ways to the end-user. Notification messages detailing the current cost and potential savings can be shown to the user, whilst real-time feedback is given through a physical prototype system, which allows users to instantly see how much energy they are consuming within different rooms in the house.

An administrative component is provided which allows carers or family members to configure the system and provide guidance on what levels of energy expenditure the older user should use on average. This information is used to inform the older user that targets are not being met and to keep the carer or family member informed and involved in the care of the user. Additionally, these set targets also allow carers/family members to closely monitor that the older user is using a minimum threshold amount of energy, thereby ensuring that they are not suffering from the lack of heating or cooling.

V. USER INTERACTION

Compounding the problem of fuel poverty, older users also suffer from technology acceptance. This adds additional constraints in which electronic systems subsumed within the home impact the experience and engagement of the older user. If the technology with which the older user has to interact is complicated or non-intuitive, this can lead to users disengaging with the system put there to improve their well-being and causes an opposite effect so that the user is even more isolated and stuck within the fuel poverty trap. The issue of reducing a lack of understanding and purpose of the system [9] needs to be considered to allow the prototype to be effective with users. In turn, Leonardi [6] highlights that the interactive medium needs to take into account the "*motor and cognitive capabilities*" [6] of the older user. When considering the perceived usefulness of the system [5] to promote older users to engage and trust the system and recommendations, a clear simple interface is required. These issues are important when considering how the older person interacts with the prototype system. However, it does pose challenges when considering the acceptance of the interface as it must not rely on something that might cause confusion or anxiety when interacting with the system. Therefore, some form of mobile device or interface would pose significant challenges to older people.

Different interfaces were constructed and evaluated to determine which one offered the older user with the best way to interpret energy usage information. One method of conveying information was through a clock metaphor whdH3 indicated the general energy usage within the home. However, this did not prove to be too useful or popular with older-users.

The approach, which ended up being taken, was to provide the older user with a simple, non-technical interface which does not rely on any interactive technology (e.g., mobile phones or smart televisions) but instead uses a simple traffic light metaphor. A mock-up miniature house was provided which uses a traffic light system located in each of the rooms within the model. Instant visual feedback is provided to the older end-user by indicating their power consumption in each room. Red indicates that they are excessively using energy within that room, amber indicates they are using more than they should but there is room for improvement with a few recommendations, while green indicates that they are using energy within the guidelines, or what the carer/family member might have specified.

Figure 3 shows the model house prototype interface. Each room of the house contains a set of LED's which indicates the real-time use of energy within that room. This was found to be the more successful of the prototypes for conveying straight forward energy usage information to older users.

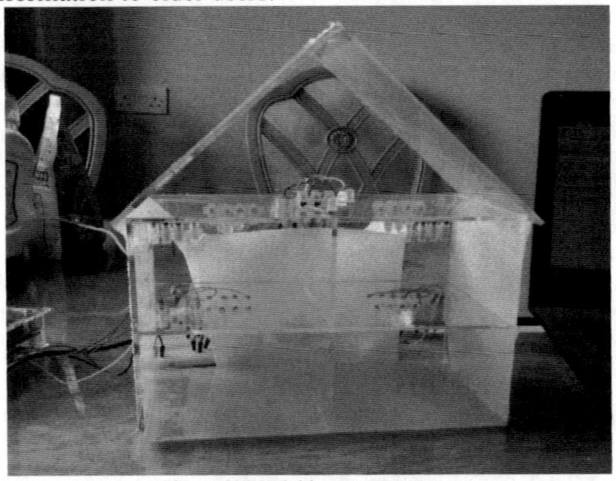

Figure 3. Model house prototype.

A. *Administrative interfaces*

The administrative interface provides carers or family members with cost saving's information either on a day-by-day basis or over a projected month. If older users felt comfortable with the technology they were also provided access to this information, rather than simply relying on the model house. For example, figure 4 shows the costs incurred for a number of monitored appliances within the home environment. Information is outlined, based on the activities that have been detected during the day and shows a comparison of their energy use. Both carers/family members and the more technically confident older user can utilise this information to determine how many times a day they perform specific activities with a view to reducing the frequency over time or to more suitable times which reduces their energy costs.

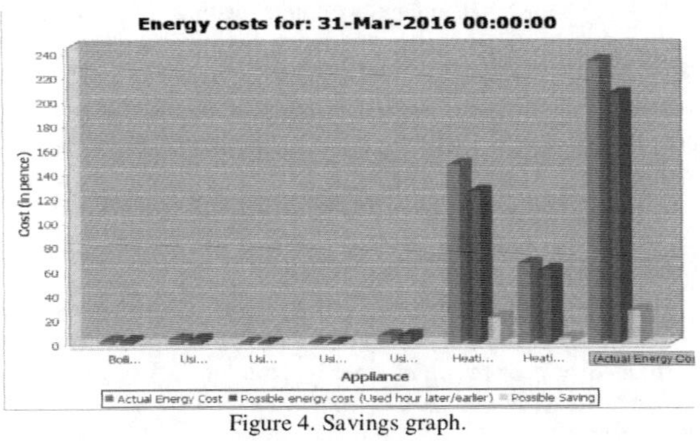

Figure 4. Savings graph.

Another way in which financial information can be conveyed to the user is through a monthly chart outlining the costs that have been incurred for each of the activities during that particular time period. Figure 5 shows how information can be presented to end-users regarding the total costs incurred and by activity.

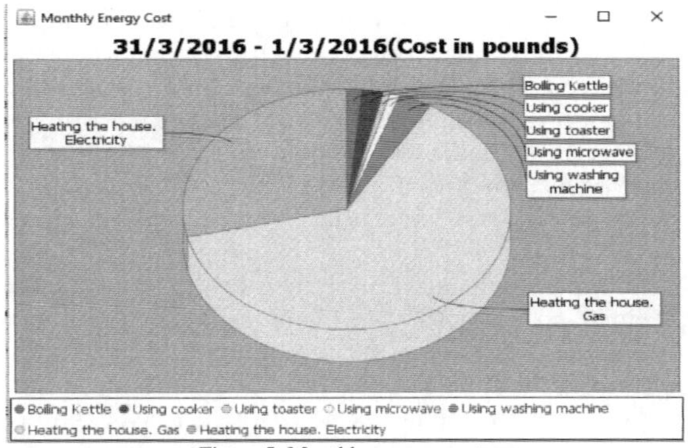

Figure 5. Monthly energy costs.

Recommendation notifications were also provided to end-users and carers/family members to allow them to determine where savings could be made by changing habits. For example, recommendations could be made to change the time of morning snacks and eating times to make best use of energy costs by moving these occurrences by a few minutes.

VI. INITIAL RESULTS

To combat fuel poverty, older users have to overcome issues with anxiety caused from the distress of lack of finance to pay for energy bills and also the fear of technology, which impedes the adoption of technology to help them become more energy efficient. Data was collected from an idealised test environment and played back through the system for evaluation purposes. The prototype was evaluated by 10 older users, whose ages ranged between 50 and 70 and were used to measure their technological fear and acceptance of using this technology. This was to determine whether monitoring energy use and recommending more efficient use of energy through modifying activity patterns, would improve their well-being by reducing their overall energy costs.

Preliminary testing has been conducted using 10 users to determine the effectiveness of the model house prototype. Older users were asked to answer a questionnaire after evaluating the usefulness of the interface. The data collected represented activities spread over several days from the test environment and was played in compressed form for them to evaluate the prototype. Activities were highlighted and explained during the playback of telemetry to help identify what was being done. 100% of the respondents identified with what the model house prototype was attempting to do by making them more aware of their energy consumption patterns. It was found that 80% of users were able to interpret the real-time information with ease while 20% of the respondents experienced difficulty at times when the system detected an increase or decrease in the energy use and notified them through flashing LED's. When addressing the question of technology acceptance and the fear of using new technology which can beset an older user, 100% of respondents said that they did not feel anxiety or fear regarding accepting information generated through the model house prototype. When asked if the prototype provided them with a way to monitor and adjust their habits during the day to make better use of energy, all respondents replied favourably (100%). In fact, it was discovered at this point that a number of users would like even more information presented to them on improving their energy expenditure. This proved to be encouraging and a validation that the model house improved their energy awareness and reduced their anxiety from worrying about fuel costs. Another group of 10 users (aged between 20 and 50) were used to evaluate the usability and functionality of the backend functionality of the system. This type of functionality would be used by carers or family members to help advise the older end-user on how to improve their energy-use by exposing them to fine-grained data regarding which activities were done and how potential savings could be made. The type of questions asked to these participants were predominantly focused on activity management, savings information, monthly projections and recommendations. The results from this evaluation showed that 100% of participants understood what the activities were and how they related to the system and data while 70% of participants were able to deal with, and manage, activity related features. The remaining 30% required extra guidance before they felt fully comfortable with the back-end system. When addressing the usefulness of the recommendations and potential financial savings, all participants agreed that the system was easy to use and offered important and helpful information on how to improve energy usage.

VII. CONCLUSION AND FUTURE WORK

Our intention was to produce a simple ICT system which would attempt to reduce the anxiety that older people experience in relation to fuel poverty as well as address the issue of accepting technology. It was found that the prototype system promoted the perceived usefulness of the system to older users. This is in addition to other areas of our research into energy-aware programming languages and runtime systems.

By focusing on the energy-awareness of everyday use from the start, the prototype system was able to successfully improve the energy use of a user over the initial trial period. The behavioural analysis and recommender systems were found to accurately identify activities which in turn aided in recognising and highlighting the cost awareness of what users were doing in association to their activities. By increasing this cost awareness of their actions, users were found to be more aware of how to reduce their energy costs by changing times or common activities to more cost effective times of the day. Carers/family members would be able to help by raising awareness of what the older user was doing on a day-to-day basis by showing them how simple changes to their daily patterns could result in tangible cost savings as well as improved health.

Following on from our preliminary results and prototype system we intend to expand trials for longer periods of time as well as integrating other technology dealing with energy-conservation within IoT devices populating these homes. Experiments on how the system affects the overall well-being of the older person will also be conducted as well as the impact of gender, disability (both physical and cognitive related), living status and age on their acceptance and use of the technology.

REFERENCES

[1] Department of Energy & Climate Change, "Fuel Poverty Report 2015." Available at: https://www.gov.uk/government/uploads/system/uploads/attachment_data/file/468011/Fuel_Poverty_Report_2015.pdf [retrieved: November, 2016]

[2] G. Flood, (2013). "British government's 'Digital By Default' plans challenged", Information Weekly (Online), available at: http://search.proquest.com/docview/1319367398?accountid=14693 [retrieved: November, 2016]

[3] The Poverty site, "UK: excess winter deaths", The Poverty Site. Available at: http://www.poverty.org.uk /67/index.shtml [retrieved: November, 2016]

[4] Greenseniors.eu, "Green Seniors In The European Union | Ageing Is Nothing For Cowards (May West)". Available at: http://www.greenseniors.eu [retrieved: November, 2016]

[5] F. Davis, "Perceived Usefulness, Perceived Ease of Use, and User Acceptance of Information Technology", MIS Quarterly, vol. 13, no. 3, pp. 319, 1989.

[6] C. Leonardi, C. Mennecozzi, E. Not, F. Pianesi, and M. Zancanaro, "Designing a familiar technology for elderly people", Gerontechnology, vol. 7, no. 2, 2008.

[7] M. Farquhar, "Elderly people's definitions of quality of life", Social Science & Medicine, vol. 41, no. 10, pp. 1439-1446, 1995.

[8] E. De Beurs, A. Beekman, A. Van Balkom, D. Deeg, and W. Van Tilburg, "Consequences Of Anxiety In Older Persons: Its Effect On Disability, Well-Being And Use Of Health Services", Psychological Medicine, vol. 29, no. 3, pp. 583-593, 1999.

[9] R. Ellis, J. Allaire, "Modeling Computer Interest in Older Adults: The Role of Age, Education", Computer Knowledge, and Computer Anxiety. Human factors, vol. 41, no. 3, pp. 345-355, 1999.

[10] Hivehome.com. "Hive", Available at: https://www.hive home .com, [retrieved: November, 2016]

[11] IFTTT.com, IFTTT", Available at: https://ifttt.com/ [retrieved: November, 2016]

[12] Honeywelluk.com, "evohome - Honeywell UK Heating Controls", Available at: http://www.honeywelluk.com/products /Underfloor-Heating/evohome-Main/ [retrieved: November, 2016]

[13] Heat Genius, "Heat Genius remote heating control benefits: save money, save energy, save time", Available at: https://www.heatgenius.co.uk/benefits/ [retrieved: November, 2016]

[14] J. Kim, H. Choi, H. Wang, N. Agoulmine, J. Deerv, and W. Hong, "POSTECH's U-Health Smart Home for elderly monitoring and support", IEEE International Symposium on a World of Wireless Mobile and Multimedia Networks (WoWMoM), pp. 1-6, 2010.

[15] T. Hargreaves, M. Nye, and J. Burgess, "Making energy visible: A qualitative field study of how householders interact with feedback from smart energy monitors", Energy Policy, Volume 38, Issue 10, pp. 6111-6119, ISSN 0301-4215, October 2010.

Tele-Rehabilitation Platform for Upper and Lower Limb in Elderly Patients, the HEAD Project.

Franco Molteni*, Cristina Gramigna*, Samuela Canobbio*, Milena Peverelli*, Silvia Aggujaro*, Mauro Rossini*, Giovanna Palumbo*, Giandomenico Caruso†, Mario Covarrubias†, and *HEAD consortium*
*Valduce Hospital, Villa Beretta, Rehabilitation Centre, Costa Masnaga, Lecco, Italy.
†Politecnico di Milano, Department of Mechanical Engineering. Via G. la Masa 1, 20156, Milano, Italy
Email: mario.covarrubias@polimi.it

Abstract—**Tele-rehabilitation systems have been increasingly studied, becoming an important complement to traditional therapy as they can provide high-intensity, repetitive and interactive treatments of the injured extremities. Several systems have been developed in research projects and some of these have become products mainly for being used at hospitals and care centers. After the initial rehabilitation performed at rehabilitation centers, patients are obliged to go to the centers, with many consequences, as costs, loss of time, discomfort and demotivation. However, it has been demonstrated that patients recovering at home heal faster because surrounded by the love of their relatives and with the community support. So, there is a strong need for rehabilitation systems to use at home. There are already some devices available on the market for home care: their main weakness is the limited strategy for motor learning. The ambition of the Human Empowerment Aging and Disability (HEAD) project is to combine assistance with appropriate feedback to close the loop in motor learning strategy for the home. HEAD platform challenging objective is to develop a novel tele-rehabilitation system and related provision services for a holistic rehabilitation aiming at restoring limb motor functional abilities, and at launching both system and services at home. The tele-rehabilitation system is based on the use of low cost sensors, connected with a gaming module for cognitive-motor rehabilitation, and integrated with an infrastructure connecting the patient at home with therapists at hospitals who monitor the rehabilitation exercises.**

Keywords–Tele-rehabilitation; Gaming; LeapMotion; Kinect.

I. INTRODUCTION

According to the World Health Organization (WHO), within the context of the action plan 2014-2021 entitled 'Better health for persons with disabilities', currently there are 150 million adults, i.e., 1 out 7, worldwide who are experiencing significant difficulties functioning. Not only, it is foreseen that the disability prevalence will increase more and more. As stated by the EC communication on Disability Action Plan (2006 2007), ageing is strictly interconnected with disability prevalence. 30% of people aged between 55-64 suffer of some form of disability, and 63% of the people with special needs are older than 45. According to the European Disability Forum (EDF), there are 80 million Europeans with disabilities. This is over 15% of the whole population. According to the European DG Health and Consumers, ageing is one of the greatest social and economic challenges of the 21st century for European societies. It will affect all EU countries and most policy areas. Let us consider that according to The World Bank Database, today the European population aged 65 or over is 18%, (90 M people), and it is expected that by 2025, more than 20% of Europeans will be 65 or over, with a rapid increase in numbers of over-80s. Unfortunately, ageing is also related to an increase of diseases and disabilities. Strictly related to the ageing of the population, stroke is the most common cause of adult disability in Europe. Currently, the European incidence is equal to 2 every 1000 persons (that is, around 1M people), but also in this case, it is foreseen that this number will double in the next 50 years. Due to the improvement on the medical stroke treatment, the 75% of people who have been affected by a stroke can survive, however about the 50% of those survivors suffer from serious disabilities, thus being not able anymore to live independently at home. It is, therefore, possible to state that with a population that more and more gets old, thus encountering more disabilities, there is the need to find new solutions for improving the quality of life of these people, by helping them in recovering their motor and cognitive skills. Today, approximately half of stroke survivors access some form of rehabilitation on discharge from acute services. Moreover, it has to be taken into account that the rate of people with disabilities or long-term health problems all over the EU Member States is estimated at 15.7% of the working-age population (around 49 Millions of the EU's overall working-age population). Thus, disability is one of the main challenges to be faced at different levels (EU Communication European Disability Strategy 2010-2020: A Renewed Commitment to a Barrier-Free Europe).

In order to improve the quality of life of the people with special needs, the WHO has drafted some recommendations: - make health care affordable; - invest in specific services such as rehabilitation. Specifically, the ambition of HEAD Platform is to strengthen and extend rehabilitation services, including community-based rehabilitation, and assistive technology. Because patients with disabilities have different healthcare requirements, the HEAD platform will be adapted to patients needs so as to provide adequate care and remain financially sustainable. There are also reasons related to the social rights, which urgently require EU to address rehabilitation issues. A study by the European Foundation for the Improvement of Living and Working Conditions has highlighted that there are ten risk factors for social exclusion. Among them we have: the disability, long-term unemployment, low quality employment or absence of employment record, health. Moreover, people with disabilities have often to stay in hospitals or residential houses, in order to be able to perform the rehabilitation. In Europe, hundreds of thousands of people with disabilities, mental health problems, older people live in large institutions. There they are alone, not connected with the society. And also, according to the European Disability Forum, more than 200.000 disabled persons in Europe are forced to live in closed institutions deprived of the most fundamental social rights. The Europe 2020 strategy has set ambitious targets for inclusive growth requiring action to be taken to promote integration and adequate livelihood of people with special needs. The HEAD results will be fundamental to contribute to the enhancement of the possibilities of people with disability to do the rehabilitation at home, surrounded by the love of their relatives, and also with the support of the community where they live.

A. *Main ideas, models and assumptions involved*

On the basis of the considerations discussed in the introduction, it follows that the society needs new rehabilitation systems dedicated to upper and lower limbs with additional capabilities, features and related services that have a more holistic approach to rehabilitation:

- offering a more effective and customized therapy for the rehabilitation of upper and lower limbs, which address cognitive and motor deficiencies, allowing for an easy customization and definition of exercises tailored on specific patients and specific rehabilitation targets, enhancing human-computer interaction through gaming, and engaging the patients in a more motivated and long-term rehabilitation program by focusing on complex, global and rewarding exercises;
- offering the possibility of performing rehabilitation programs in autonomy, without the need of the continuous presence of therapists, but under their continuous monitoring, and which can be performed quietly at home;
- allowing therapists to use a complete and rich set of data recorded from the exercises executed by the patients to improve knowledge on rehabilitation and therapies.

B. *The innovation potential of HEAD Platform*

The innovation potential of the HEAD Platform is:

- at technological level: novelty of the integration of low cost tracking devices (kinect, LeapMotion, etc.), combined with games, in the rehabilitation sector;
- at rehabilitation therapy level: more effective therapy that is customised to patients, motivating, rewarding, monitored, plus a novel knowledge based system for therapists and physicians;
- at socio-economic level: better quality of life for impaired patients and their family, thanks to a better rehabilitation service, and decrease of social costs of the rehabilitation practices; better exploitation of therapists skills and time, which also means an increased number of patients they are able to assist.

The HEAD Platform system features have an innovation potential on the kind of rehabilitation therapy that it will be possible to perform. Patients will be able to make the rehabilitation at home, on a regular basis, enjoying themselves and with more motivation and consciousness of the improvements, and being remotely monitored by the therapists. This will have an important effect on the patients. It is foreseen that patients will be able to reacquire the lost motor skills in shorter time because they can practice more often, for longer time, executing dedicated and customised exercises. Besides, it is expected a lower abandonment of the therapy, as it will be more pleasant and motivating. Extremely important is the fact that the rehabilitation will be performed in a family context. It has been demonstrated that patients recovering at home after a period spent in hospital, being with their loved ones and with their belongings, heal faster. So, being at home is very important from the psychological point of view, and this impacts on the quality and time of the recovery. Therefore, we can say that the innovation potential of HEAD Platform will bring benefits for patients, and consequently, there will be also important effects on society and societal costs. First, family members won't be obliged any more to absent work to accompany the patients to care centers. Then, if the skills will be reacquired more rapidly and completely (or very close to) the persons can improve their general quality of life, reacquire their independence and eventually can be re-integrated in the workplace.

C. *Main outcomes*

Therefore, HEAD Platform aims at addressing rehabilitation problems and shortcomings as listed in the Introduction:

1) Patients will be more motivated to do therapy following the prescription of the therapist thanks to the playful and entertainment character of the system. The hospitals participating to the project expect that along the rehabilitation treatment at least 60% of patients perform therapy in a more exactly and individual adapted way than now (with habitual rehabilitation treatment), and also that after the neuromotor stabilization rehabilitation (when patients usually are discharged from rehabilitation centers), 70% of patients will continue rehabilitation activities at home, on a regular basis. This would be a higher percentage compared to the present figure of 10%.
2) The clinic partners make the assumption that patients will be more motivated to perform the exercises when using the HEAD platform, so they are expected to exercising on a regular basis as prescribed by therapists, and also that the system will allow therapists to customize exercises to be more focused to the specific psycho-motor problems of a single patient. Consequently, it is expected that the recovery of the skills that one can get is in a shorter time compared to those obtained through the rehabilitative practice currently in use. The estimate is that the time is shortened by 20%.
3) As a result of the fact that the patient will be more relaxed during the sessions of rehabilitation, which can be made using a modality entertainment and even at home, the patient's psychological state and well-being will certainly increase considerably. In order to assess if the HEAD platform will definitively improve the general well-being of the patients, a set of 'quality life questionnaire' will be prepared. It will be administered to the patients at the beginning of the rehabilitation treatment, at an intermediate period (according to the length of the rehabilitation treatment) and at the end. The questionnaire will be also administered to the caregivers of the patients, so as to gather a larger perspective on the issue.

The paper is organized as follows. Section 2 presents an overview of related about gesture interaction systems. Section 3 presents the HEAD Platform architecture. Section 4 describes one example of the HEAD rehabilitation exercise. Section 5 describes user study approach and its preliminary validation. Finally, Section 6 draws some discussion and conclusions.

II. GESTURE INTERACTION

We have developed a visualization and interaction system integrated with the gaming module that visually renders the virtual environment for rehabilitation and also the user's virtual hands and body in real-time. This functionality requires tracking the user's hands and body in the physical space. In order to provide a realistic immersiveness, the tracking and the representation of the user's hands and body in the virtual environment should be accurate and timely. In fact, some studies have shown that if the users are able to see the virtual rendering of their hands and legs and their movements relative to the movements of other objects there is a much better chance that they will feel that the virtual hands embodies their intentions and actions [1].

In designing the system, we have considered the new generation of so-called 'natural user interfaces' (NUI) technologies [2], which track the user's hands, fingers, or entire body in 3D, without wearing any kind of invasive device. Free-hand gestures can provide effective and natural methods for 3D interaction with virtual shapes, which can provide fluidity in the interaction.

Traditional haptic systems [3][4] and desktop interaction approaches, typically based on devices, such as keyboard and mouse and touch pad interfaces, are often designed for 2D interfaces and consequently are less effective and usable for 3D interaction [5]. In addition, many of the techniques that are used for 3D interaction require the user to wear or hold 3D tracking devices and also require several markers attached to the user's hand or body. The use of markers can make the system more difficult to configure and is inappropriate for some scenarios. Reasonably precise 3D sensing techniques, which can recognize freehand movements, are now available at low cost (e.g., Microsoft Kinect, Asus Xtion, and Leap Motion). These types of devices do not require on-body attachments or hands-on tracked devices, thus enabling very low configuration interaction. In this way, users can interact with the system naturally through their hands or body movements, without using complex commands.

However, simply because the interface is based on 'natural' gestures, this does not reduce the need for careful interface design [8].

Gestural interaction techniques have been investigated for a long time and different gesture types have been designed and evaluated. For example, for the selection of occluded targets on small touch screens using a finger [9]. Gestural selection has been explored in order to select distant targets on large displays [10]. Early freehand interaction systems needed fiducial markers on the users or data gloves in order to track the user's gestures. For example, the authors in [11] used the movements of fingers to simulate "mouse-clicking" and also to investigate freehand gestural interaction with ambient displays [12] developing design principles and an interaction framework for interactive public ambient displays. Bimanual marking menu selection [13] used a kinect system at clese range to track the finger's pose and movement to select from a marking menu. However, this method requires setting up the camera under the desktop at a specific angle, so losing the convenience of free hand interaction. In [14] is presented a Kinect One sensor-based protocol for the evaluation of the motor-performances of the upper limb of neurological patients during rehabilitative sessions. Freehand gestural input has also been explored for virtual object manipulation [15] including on curved surfaces [16] and projected directly on to everyday objets [17]. In this research, it is illustrated the importance and immediacy of freehand gestural interaction in daily life use cases. Participants felt that freehand pointing is intuitive but needs more precise operation. There is a lot of previous research on object and option selection in both 2D and 3D interfaces, however, most previous research used hand-held tracked devices or fiducial markers to enable camera-based tracking [5].

III. THE HEAD PLATFORM

The HEAD platform allows us to implement a set of exercises for upper-limb and lower-limb rehabilitation. The interaction with virtual objects is performed through hand and body movements and gestures. The most innovative aspect of the platform is that the interaction is performed by using the LeapMotion and the Kinect sensors in a unique platform. The following section describes the HEAD platform components in detail.

A. Hardware components and software modules

The system is easily customizable and provides a simple environment for project deployment to use with multiple platforms, with no need for additional configuration. The HEAD platform consists of the following hardware components:

- a computer;
- Leap Motion controller, used to track the users' hand movements and to detect gestures;
- Kinect controller, used to track the users' body movements and to detect gestures;
- FITBIT bracelet which measures the amount of physical activity.
- PC speakers or headphones, used to render sounds.

For what concerns the software modules, after comparing several gaming engines, Unity3D R5 environment [6] has been selected for the implementation of the visualization and interaction. The main reason for selecting Unity3D is that it has a powerful interface that allows visual object placement and property modification during the interaction. In particular, the required features of Unity3D for the development of the games application are as follows: 3D rendering of virtual shapes; Graphical User Interface (GUI); Physics engine (used for handling collision detection); Collision detection; Integration with Leap Motion and KINECT (used for user's gesture recognition).

B. Head Platform Architecture

The HEAD platform is used by the patients, by the neurologist in charge of the patient and by other clinical specialists (e.g., by the neuropsychologists and physiotherapists). Each of the users has proper functions: the patient can see the daily exercises therapy, the neurologist can assign, modify or evaluate the therapeutic plans, the medical personnel can check how the rehabilitation is going. The main characteristic of the HEAD platform are related to the possibility for the clinicians to assign rehabilitation plans to be performed at home, and the holistic approach to rehabilitation, as the plan includes physical, cognitive and behavioural therapies/exercises. The interaction of the different users and the logical architecture of the platform are showed in Figure 1.

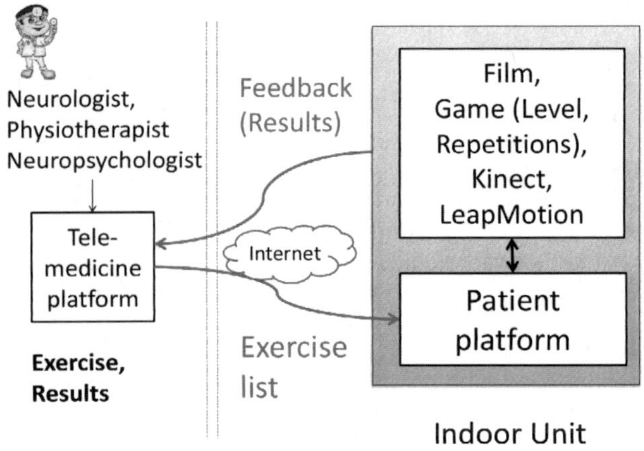

Figure 1. Head Platform Architecture.

The Indoor Unity represents the patient's interaction with the platform. The patient might be equipped with different devices that can be provided with the HEAD platform: a dedicated PC, used by the patient to see the daily activity exercises; an activity tracker (FITIBIT©) to measure the amount of physical activity of the patient. The platform foresees also other devices, which are used only at the hospital (e.g., an EEG portable helmet to measure variations in the brain activity of the patients during specific cognitive or attentive tests). All the patients data is stored in a clinical based certified database which is managed by TELBIOS [7] a local industrial partner of the HEAD Project.

IV. HEAD Rehabilitation Exercises

The HEAD exercise database consists of three different categories which are connected through a film database and a picture database. A film is used in order to catch the attention of the patient.

- **Category 1:** Film $\Longrightarrow$ Exercise $\Longrightarrow$ Film: All the exercises which are inside this category starts with the visualization of a film, after a few seconds the film stops and the rehabilitation exercise starts using a picture keyframe of the film. Once the rehabilitation exercise has been completed, the film is enabled and the patient can see again the next part of the film. After few seconds the film stops again and the rehabilitation exercise starts again, and so on. This category has been designed for motor-limb rehabilitation.
- **Category 2:** Film $\Longrightarrow$ Exercise: In this category, the exercises have been designed in order to start with a film without stopping or pausing the film visualization. The rehabilitation exercise starts once the film is completed. This category of exercises allows neuro-motor rehabilitation.
- **Category 3:** Exercise $\Longrightarrow$ Film: This category of exercises starts directly with the rehabilitation exercise and the video is seen by the patient only if the exercise has been completed. The film is used as an award.

V. Users Study

A tele-rehabilitation system should be validated in order to demonstrate its usefulness. Before a rehabilitation environment can be used in a rehabilitation protocol it is necessary to perform several tests and see if the proposed exercises are usable and potentially effective.

In our research, a user study has been carried out in order to test the tele-rehabilitation platform initially with healthy people. In particular, we were interested in checking if any issues may arise during the use of the HEAD application, in particular any issues related to gesture interactions.

A. *User study with healthy subjects*

10 healthy users, 6 female and 4 male, aged between 18 and 23 participated to the preliminary tests. Before the test, the participants were asked to fill in a pre-test questionnaire with their data, confidence to use games and also hand gestural technologies. In addition, we asked the participants to compile a symptoms check-list related to the sense of sight.

Figure 2-a shows part of the HEAD platform in which the clinical team is able to select the different parameters for the rehabilitation exercise. It is possible to select the video, the level, the device and the arm/hand side which will be involved in the exercise therapy.

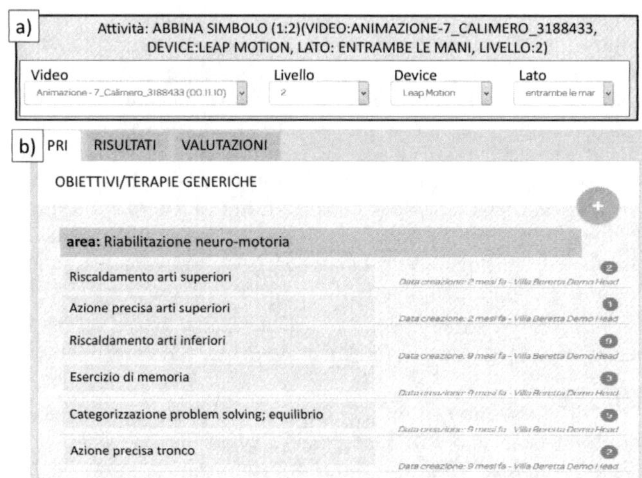

Figure 2. HEAD Platform for the clinical side.

Figure 2-b shows the general therapy and all the different rehabilitation areas.

Participants were instructed about the task, and they were allowed to use the hand gestural approach until they felt sufficiently confident. The task consisted in using the '**match**

the symbols' exercise. Two tests were performed, in different conditions.

First test. The participants were asked to perform the '**match the symbols**' exercise and to use the non dominant hand to interact with the virtual objects. This was made to introduce in the users some difficulties typical of patients with upper limb disabilities.

Second test. The participants were asked to complete again the task using the dominant hand.

The two tests lasted approximately 8 minutes (4 minutes per each test). Figure 3-a shows the HEAD platform once is started by the patient and after performed the login via user-name and password. The patient selects the rehabilitation exercise, in this case the '**match the symbols**', which consist in matching the three similar figures (Figure 3-b). The system counts as an error if the patient does not match properly the figures. If the three figures are matched accordingly, the main figure changes its color to green and the patient is able to grab the three pictures group in to the blue container (Figure 3-c). Once all the pictures groups are matched the patient is able to see the film as an award. When the film finished the system shows the final score in form of stars (Figure 3-d). Then, the results are immediately stored to the collector database.

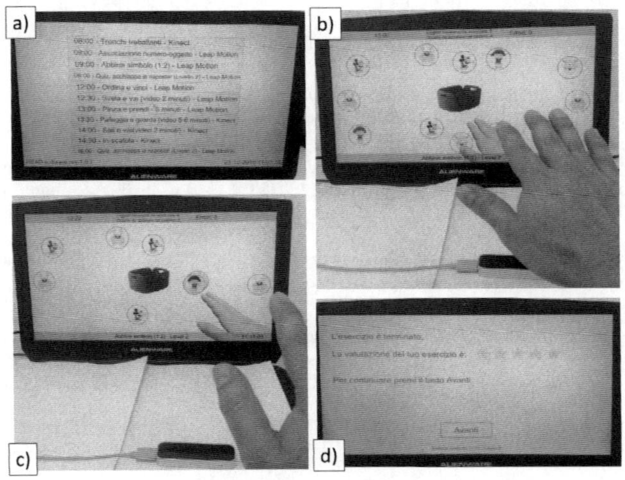

Figure 3. '**Match the symbols** exercise'.

After completing the two tests, participants were asked to fill in a post-test questionnaire aimed at evaluating the following aspects:

- general impression of the exercise, including the proposed concept, the system as a whole, the willingness to use it again, its easiness of use;
- knowledge acquired in using the scenario, including the easiness when started using the rehabilitation exercise, the effectiveness of the information displayed, and the willingness to use the system again in the future;
- evaluation of the system functionalities, including the quality of the graphic user interface, the perceived quality of the exercise, and sound feedbacks, and finally the ability of the hand/finger interaction modality.

The questionnaire was organized in a 6 points Likert-scale, from 1 (which is the most negative value) to 6 (which is the most positive value).

B. Analysis of the results

The charts on Figure 4 show the results of the questionnaires.

The HEAD application achieved a high evaluation rate relatively to the aspects concerning the system in general, a quite positive evaluation of the easiness in using it as a whole (chart **a** in Figure 4).

The *knowledge acquisition* section of the test intended to go more in details in understanding and evaluating the system from the user's perspective. Chart **b** in Figure 4 shows the results. Overall, the collected data show a positive evaluation. Only one user assigned a very low rate for what concerned the easiness of using the system the first time. But the same user was convinced that the following time it would have been easier and more natural to use it.

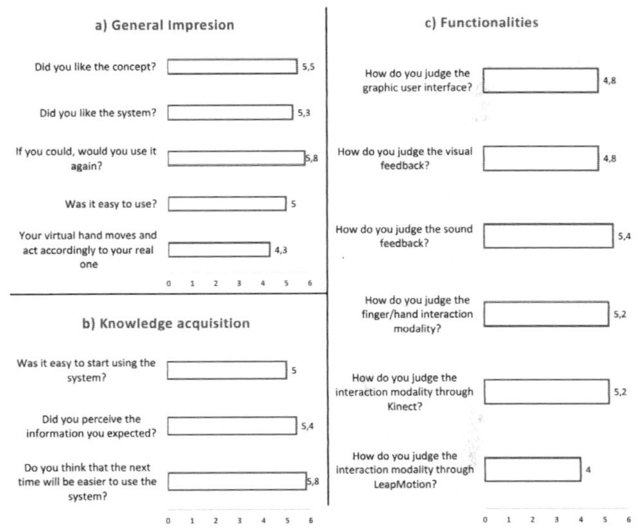

Figure 4. Results of tests performed by the healthy subjects.

It is possible to observe from chart **c** in Figure 4, that the participants have judged interesting the gesture interaction modalities proposed (finger/hand gestures) through the Leap-Motion and Kinect sensors. A positive evaluation has been given also to the visual and sound feedbacks which were coherent with the user's interaction.

All the objects (images and films) in the exercise are clearly recognised by the participants. All participants agreed to feel more confident in using the hand gesture interaction system towards the end of the test than at the beginning, and they felt to be able to improve their performance after using it for a longer period.

Five participants reported some incongruence between their real hand movements and the virtual hand avatar. This is due the Leap Motion controller, which sometimes produces anomaly reading data, such as reporting identical position although the finger had moved, or reporting false positions far away from the actual fingers [18]. These anomalies however, were usually short-termed and did not represent a significant impact in the user's performance, and in general in the overall results. Nevertheless, these anomalies were known to happen [19] and therefore expected. This preliminary study was designed to cope with these issues.

Additionally, in this study we evaluated only the process of selection of the objects (e.g., images through the gesture interaction) in the rehabilitation exercise. These two conditions were performed separately. In general, more sophisticated actions can be performed while interacting with a different rehabilitation exercise, such as advanced manipulation (e.g., changing position and orientation of virtual objects, activating some interactions through the pinch gestures), which will be considered in future works.

VI. Discussion and Conclusion

The use of tele-rehabiliattion systems for upper and lower limb rehabilitation has been demonstrated in literature to be a valid approach.

In this view, the paper describes a tele-rehabilitation based on hand and body gesture interaction. We have performed some preliminary tests in order to prove the HEAD approach. The preliminary test results reported in the paper are positive for what concerns the quality of the hand and body gesture interaction while executing the rehabilitation exercise. On this basis, we conclude that the gestural interaction system provides users with an effective and natural method to interact with the rehabilitation exercises.

The clinical protocol implies that each patient is able to use the HEAD platform 12 times in the rehabilitation center and 60 times at home. The total patients involved in the HEAD project are 100. All of them are using the HEAD platform in the rehabilitation center and at least 30 at home.

Acknowledgment

This research has been partially developed within the context of the project HEAD: Human Empowerment Aging and Disability, Funded by CARIPLO Foundation. The authors are grateful to the HEAD Consortium for the unconditional support for the HEAD Project.

References

[1] C. Cameron et al. "Hand tracking and visualization in a virtual reality simulation," April 2011, pp. 127–132.

[2] D. Wigdor and D. Wixon, "Brave NUI World: Designing Natural User Interfaces for Touch and Gesture," 1st ed. San Francisco, CA, USA: Morgan Kaufmann Publishers Inc., 2011.

[3] M. Covarrubias, M. Rossini, G. Caruso, G. Samali, C. Giovanzana, F. Molteni, M. Bordegoni, "Sound Feedback Assessment for Upper Limb Rehabilitation Using a Multimodal Guidance System," in International Conference on Computers Helping People with Special Needs. ICCHP 2016 - Proceedings, July 2016, pp. 529–536.

[4] M. Covarrubias, M. Bordegoni, U. Cugini, E. Gatti, "Supporting Unskilled People in Manual Tasks through Haptic-Based Guidance, Information Systems Research and Exploring Social Artifacts: Approaches and Methodologies. Edited by IGI Global, 2012. pp. 355–378.

[5] G. Ren and E. O'Neill, "3d selection with freehand gesture," Computers & Graphics, vol. 37, no. 3, 2013, pp. 101 – 120. [Online]. Available: http://www.sciencedirect.com/science/article/pii/S0097849312001823 [retrieved: February, 2017]

[6] [Online]. Available: http://http://unity3d.com [retrieved: February, 2017]

[7] [Online]. Available: http://www.telbios.com/telbiosconnect/ [retrieved: February, 2017]

[8] F. Bacim, M. Nabiyouni, and D. A. Bowman, "Slice-n-swipe: A freehand gesture user interface for 3d point cloud annotation," in 3D User Interfaces (3DUI), 2014 IEEE Symposium on, March 2014, pp. 185–186.

[9] K. Yatani, K. Partridge, M. Bern, and M. Newman, "Escape: A target selection technique using visually-cued gestures," in Conference on Human Factors in Computing Systems - Proceedings, 2008, pp. 285–294.

[10] A. Bragdon and H.-S. Ko, "Gesture select: Acquiring remote targets on large displays without pointing," Conference on Human Factors in Computing Systems - Proceedings, 2011, pp. 187–196.

[11] D. Vogel and R. Balakrishnan, "Distant freehand pointing and clicking on very large, high resolution displays," UIST: Proceedings of the Annual ACM Symposium on User Interface Softaware and Technology, 2005, pp. 33–42.

[12] R. Balakrishnan and D. Vogel, "Interactive public ambient displays: Transitioning from implicit to explicit, public to personal, interaction with multiple users," UIST: Proceedings of the Annual ACM Symposium on User Interface Softaware and Technology, 2004, pp. 137–146.

[13] F. Guimbretière and C. Nguyen, "Bimanual marking menu for near surface interactions," Conference on Human Factors in Computing Systems, 2012, pp. 825–828.

[14] M. Caimmi, A. Chiavenna, M. Malosio, L.M. Tosatti, "Kinect One-based biomechanical assessment of upper-limb performance compared to clinical scales in post-stroke patients," Proceedings of the Annual International Conference of the IEEE Engineering in Medicine and Biology Society, EMBS., 2015. pp. 5720-3.

[15] P. Song, W. Goh, W. Hutama, C.-W. Fu, and X. Liu, "A handle bar metaphor for virtual object manipulation with mid-air interaction," Conference on Human Factors in Computing Systems, 2012, pp. 1297–1306.

[16] H. Benko, R. Jota, and A. D. Wilson, "Miragetable: freehand interaction on a projected augmented reality tabletop," Proc. of CHI'12, 2012, pp. 199–208.

[17] C. Harrison, H. Benko, and A. Wilson, "Omnitouch: Wearable multitouch interaction everywhere," UIST'11 - Proceedings of the 24th Annual ACM Symposium on User Interface Software and Technology, 2011, pp. 441–450.

[18] V. K. Adhikarla, J. Sodnik, P. Szolgay, and G. Jakus, "Exploring direct 3d interaction for full horizontal parallax light field displays using leap motion controller," Sensors, vol. 15, 2015, pp. 8642–8663.

[19] J. Guna, G. Jakus, M. Pogacnik, S. Tomazic, and J. Sodnik, "An analisis of the precision and reliability of the leap motion sensor and its suitability for static and diynamic tracking," Sensors, vol. 14, 2014, pp. 3702–3720.

A Novel Kinematic Model for Wearable Gait Analysis

Marcello Fusca, Paolo Perego, Giuseppe Andreoni
Design Dept. – E4Sport Lab
Politecnico di Milano
Milan, Italy
e-mail marcello.fusca@polimi.it

***Abstract*— The work studies the acceleration of a single tri-axial accelerometer fixed at the sacrum position on the back of subject next to the COM (Center of Mass). The correlation between the cycle of the COM and the cycle of the walking is analyzed by using a harmonic oscillator as a model for human locomotion. The COM position is calculated without any integration of its acceleration. A double integration of raw accelerometer data can result in an accumulation of drift error resulting in wrong position and distance or step length evaluation. The acceleration of a harmonic oscillator is directly proportional to the position. We evaluate the COM position and translation into a sinusoidal pattern. For every step cycle, the maximum of COM amplitude is used to give the relative S step length. This kinematic model generates the properly attended values; all steps are detected and the absolute accuracy error in the measurement of the step length, ranges from 0.32% to 3.33% with a mean value 2.17%. In the model, many output parameters are processed to study the subject movement analysis, but all that parameters should be compared with the gold standard values using appropriate protocols. We use data of Swedish adult people to obtain coefficients C to evaluate S_{mean} (mean anthropometric step length). S_{mean} is used only as a reference, but is not the S value measured by the model. New protocols and data verification are carried out. The expectation is to develop a dedicated tool to support diagnosis and rehabilitation.**

Keywords-wearable device; wearable sensors; gait analysis; human kinematics; clinical application; algorithms.

I. Introduction

Gait analysis is a complex and expensive technology. The setup of a limited working area in laboratory, the use of markers on the subject and the data analysis are not simple to approach. Outside the laboratory, the use of the system is not artless. Healthcare requires a novel ecologic approach to the movement analysis in order to make it friendly and designed under ergonomic constraints as well as the performance assessment in agonistic sports and/or clinical follow up in home monitoring. The wearable sensors are a possible solution to this problem. They are easy to use and not intrusive, so that it is possible to monitor subjects everywhere [1][2]. The introduction of smart fabrics and wearable sensors improves and simplifies the development of these sensors, the evaluation of the movements and also the rehabilitation of patients in their clinical pathway. In fact, the possibility to embed sensors directly into the user's garments becomes real [3][4]. In this way, their use allows for a natural walk while monitoring is in progress. Clinical tests are often conducted with manual counting of times, steps and distances; their confirmation is carried out through the support of concurrent video analysis. The use of wearable devices for the gait analysis without optoelectronic analysis is possible. This matter is still under investigation. The need to monitor health status of patients drives an improvement and an evolution to use remote control systems for analyzing data through a trained medical center. Transmission and storage of clinical data is driven by attention to the security and privacy [1][5]. On the research side, more accurate biomechanical models are being implemented, the improvement of signal processing and advanced analysis algorithms are a focus of development so to enhance the interpretation of the output data of wearable sensors for decision making [2][6][7]. Every new wearable analysis tool for gait analysis is a strong target; better results need to be introduced into the clinical practice in order to exploit for example a wearable 6MWT (Six Minutes Walking Test), or a wearable TUG (Timed Up and Go), or other trials, without the particular limitations of a laboratory. Our work goes into this direction: we developed a novel kinematic model and related method to process 3D inertial accelerometer data and to compute parameters of gait analysis. The structure of the paper is as follows: section 2.1 the Biomechanical model; section 2.2 the experimental setup; sections 2.3 and 2.4 the Methods A and B to analyze raw data; sections 3.1 and 3.2 the Data Analysis with Method A and B; section 4 the conclusions.

II. Materials and Methods

This section reports the description of the methodology used for the kinematic model. We divided the section in four subsections in which we explain the Biomechanical model, the experimental setup, the processing methods for bot the parts.

2.1 Biomechanical Model

Describing the movement of a subject, we have to consider the external forces and, therefore, the accelerations acting on his/her body. In absence of other forces, we always have the action of the gravity force, and then the subject has to produce a counterbalanced force to remain in equilibrium even when stationary. The COM (Center of Mass) is a single point where we can think that the whole mass of the body is concentrated, so that it is equivalent to the entire considered object, where the external forces act according to the Newton's laws of motion. In a standing posture, its position is typically about 10 cm lower than the navel, in the sagittal plane and in correspondence of the anterior superior iliac crests (the top of the hip bones). To know as the COM moves, it means to know how the object moves. For this reason, we studied human walking by evaluating the acceleration signals of a single tri-axial accelerometer fixed on the pelvis of subjects next to the COM, i.e. in correspondence of the second sacral vertebra.

In the FIGURE 1 a subject walk and the positions in time of the COM are presented with circles and a trajectory. L is the leg length, θ is the hip angle in the sagittal plane and S is the step length. While the subject moves a step S, the COM moves vertically along an oscillating path; the maximum oscillation amplitude is hCOM. When the subject moves the next steps, the cycle repeats. While walking, the COM oscillation pattern is considered sinusoidal in the vertical and mediolateral directions [8]. In this work we describe the COM vertical oscillation, but the same method is applicable to the COM mediolateral oscillation. To describe human locomotion, it is analyzed the correlation between the cycle of COM and the cycle of walking, by using a harmonic oscillator model. The legs are considered rigid bodies. The swinging movement is described by a pendulum model [9].

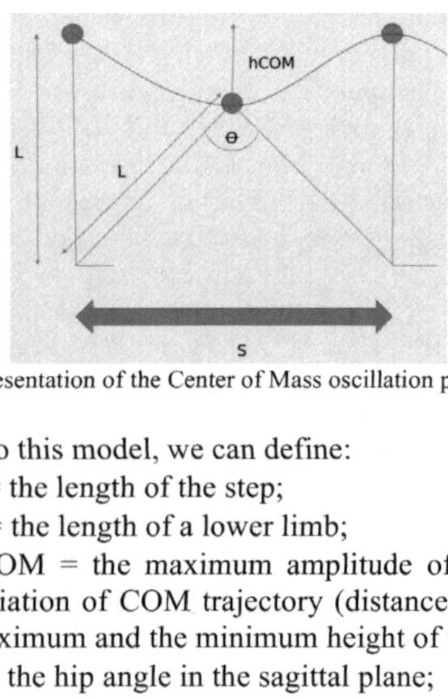

Figure 1. Representation of the Center of Mass oscillation path during walk

According to this model, we can define:

- S = the length of the step;
- L = the length of a lower limb;
- hCOM = the maximum amplitude of the vertical variation of COM trajectory (distance between the maximum and the minimum height of the COM);
- θ = the hip angle in the sagittal plane;

$$S = 2 * \sqrt{2 * L * hCOM - hCOM^2}.$$

The COM position must be calculated. A direct double integration of raw accelerometer data gives the position which can result in an accumulation of drift error giving wrong position and wrong distance or wrong step length. Other studies chose to double integrate and report solution to solve the drift error [6]. We do not carry out a double integration. Some information may be obtained using the property that the acceleration of a harmonic oscillator is directly proportional to the position. For every step cycle, the maximum of COM amplitude is used to give the relative S step length and distance.

2.2 Experimental setup

We used a wearable 3D accelerometer (Protheo SXT s.r.l., Lecco, Italy) [10]. Protheo technical details are: 85 (l) x 53 (w) x 16 (h) mm of size, 70 g of weight, 4 digits LCD, on board ARM7 microprocessor and raw acceleration sampling frequency of 128 Hz. This device is able to log tri-axial accelerometer signals into the internal memory (up to three days of continuous monitoring) which can be downloaded at the end of the acquisition by Bluetooth® data transmission. The data storage allows for recording different tests in sequence. When we carry out different tests in sequence, we have no way to monitor over the data recording until the recording is downloaded. We calibrated the output raw signals when the data processing is carry out. Protheo could record also an electrocardiogram of the subject while he is making the test, but this option is not used for this study. The system is worn at sacrum position on the back of the subject, by means of an elastic band with a pocket for fixing the device firmly to the body. We recorded six independent walking tests of a single subject. In the test protocol the subject walked with shoes at self-selected speed over a linear path of 31.2 m; the walking time is not a constant. To control the test, the step length was kept fixed at 60 cm, so that 52 steps were necessary to complete the path. These values are the true imposed values used for accuracy assessment. For this purpose, a linear set of 60 cm interspaced lines was drawn to drive the position of the tip of the foot at each step. This step value is very close to the subject natural one that has the anthropometric estimated value of $S_{mean} = 61.9$ cm (mean anthropometric step length). We evaluate S_{mean} using data of Swedish adult people but different values are possible compared with different cultural background [11][12]. From this data we obtain the C coefficients depending on gender, speed and H (height) of the subject as in TABLE I.

$$S_{mean} = C * H$$

S_{mean} is a reference and S the measured step length. C permits the evaluation of the mean anthropometric step length depending on gender, heigth and speed. Gender, age, weight and following anthropometric measurements of the subject under analysis were taken to complete the biomechanical model: a) lower limb (ground-greater trochanter); b) ground-malleolus; c) lateral condyle-greater trochanter; d) malleolus-lateral condyle; e) fifth metatarsal-malleolus; f) width of the foot; g) length of the foot to the ground; h) outer distance between the feet. The subject is a healthy male (w = 80 kg, h = 181 cm, age = 51 years) with normal BMI (Body Mass Index) 24.4. TABLE II reports his anthropometric measures. In the model, we use the Tanaka formula [13] to compute the HRmax (maximum Heart Rate) in bpm unit and the HRR (Heart Rate Reserve); we need the heartbeat as an input value at the beginning of the test with the resting subject and at the end of the test to compute the cardiac effort during exercise. The energy expenditure is evaluated in [METS]. Raw accelerometer data processing was implemented in Matlab© software suite. Two different processing methods were compared (Method A and B).

TABLE I. THE C COEFFICIENT CALCULATED USING SWEDISH REFERENCE DATA FOR NORMAL SUBJECTS;

Speed	Low < 0.90 m/s	Normal [0.90-1.40] m/s	Fast >1.40 m/s
Man	0.2928	0.3422	0.3956
Woman	0.3102	0.3539	0.3994

TABLE II. ANTHROPOMETRIC MEASURES OF THE SUBJECT UNDER ANALYSIS.

Body Segment	Value [cm]	Body Segment	Value [cm]
a	96	e	14.5
b	12	f	9
c	41	g	30
d	43	h	32

2.3. The Processing Method A

The peaks in raw data were detected by peakdet.m function [14], using a 5th order pass-band Butterworth filter (band: 0.5 - 4 Hz). To identify S, we used the peak positions detected by the previous peakdet routine, applying the harmonic oscillator model to the original raw signals filtered with a low-pass 19th order Butterworth filter, with these cut-off frequencies for each acceleration:

- Antero Posterior acceleration: 6 Hz;
- Vertical acceleration: 7 Hz;
- Medio Lateral acceleration: 8 Hz.

The evaluation of hCOM was carried out by applying a cut-off threshold of 6 cm to the double of maxima amplitude of COM sinusoidal pattern. The amplitude to be considered in the model is the length between the maximum and the minimum for every COM oscillation. The time between two consecutive vertical peaks is the single step time. In order to identify the starting step (left or right), the analysis of mediolateral acceleration was made. The same analysis can be applied in order to extract the asymmetry of right and left steps [15]. The kinematic model can extract the following parameters:

- BMI [kg/m^2];
- cadence [step/min] and frequency [step/s];
- stepping time [s];
- stride's periods [s];
- right and left step's periods [s];
- displacement of COM [m];
- speed [m/s];
- speed for the right and left steps [m/s];
- right and left step lengths [m];
- length of steps [m];
- incremental distance traveled at each step [m];
- pace distances [m];
- pace angles [degrees];
- pace coefficients of walking efficiency;
- number of steps and strides;
- sagittal hip angle [degrees];
- the first right or left leg support;
- base of support both with aids that without [m^2];
- width of steps [m];
- acceleration peaks at ground support phase [g];
- step and stride indices of regularity and symmetry;
- total and incremental energy expenditure [METS];
- the power spectrum of the accelerometer signal;
- maximum heart rate and heart rate reserve;
- 6MWT predicted normal distance value [m];
- report with office format.

2.4. The Processing Method B

Method A is consistent if subject walks with a constant step, but this is not a normal constrain; if there is a velocity variation for external causes or other voluntary choices, a new set of processing filters has to be applied to the acquired data for the research of peak acceleration and the relative values. To verify this hypothesis, we carried out a set of experiments by asking the subject to walk on a treadmill and progressively increasing its speed (from 0.5 to 1.7 m/s). The same concept could be applied to pathological patients, walking slower and asymmetrically, such as stroke patients (walking speed < 0.5 m/s).

This issue was faced in the second approach here proposed (Method B), as an evolution of Method A.

The peaks of raw signals that identify steps, were detected by peakdet.m function; the low-pass filter used is different according to the walking speed:

- at high velocity, a 5th order low-pass Butterworth filter is applied with the following cut-off frequencies:
  - Antero Posterior acceleration: 1.8 Hz;
  - Vertical acceleration: 1.8 Hz;
  - Medio Lateral acceleration: 0.9 Hz;
- at low velocity, a 4th order low-pass Butterworth filter:
  - Antero Posterior acceleration: 35 Hz;
  - Vertical acceleration: 5 Hz;
  - Medio Lateral acceleration: 3 Hz.

To identify S, we used the previous peak identification, applying the harmonic oscillator model to the original raw signals filtered by a 19th order Butterworth filter with the following cut-off frequencies:

- Antero Posterior acceleration: 6 Hz;
- Vertical acceleration: 35 Hz;
- Medio Lateral acceleration: 8 Hz.

We used six different approaches to define hCOM threshold:

Mode 0: if we do not know S, the length of the expected step, hCOM threshold is estimated by the median of twice the absolute value of the COM amplitude trend; this value is multiplied by weights that depend on the average walking speed;

Mode 1: if we know the expected step S, we impose the hCOM threshold equal to the expected hCOM

$$hCOM_{expected} = L - \sqrt{L^2 - (S/2)^2}$$

Mode 2: as in mode 1, but the expected hCOM threshold is increased of the 20%;

Mode 3: the hCOM threshold is the product of hCOM value evaluated by linear interpolation and a set of weight correction factors depending on speeds, so to matching the value of COM displacement measured by Orendurff [8], with the expected COM displacement of the model;

Mode 4: the hCOM threshold is the Lulic's COM amplitude [16] with the weights of Mode 3;

Mode 5: the hCOM threshold is 0.6 m.

Through the proper choice of the mode according to subject's feature in his/her different scenarios, the model calculates the correct length values and the number of strides and steps.

III. RESULTS

The step counting obtained from the device was compared with the true reference value i.e. that one taken manually by the observing operator.

3.1 The Data Analysis with Method A

The Method A (TABLE III) detected all 52 steps in the test (1, 2, 3, 5). In the tests (4, 6), 51 steps are detected. The absolute accuracy error in the measurement of the step length, ranges from 0.37% to 10.78% with a mean value 4.94%; the absolute accuracy error in the measurement of the walking total path, ranges from 0.37% to 10.78% with a mean value 5.22%. The subject covers the path from a standing start; the first steps are not regular because the subject is still not in a steady state. If we replace the initial steps with the next ones, we have a steady state for all the path. If the high frequencies contribute in the oscillator modeling, the removal of the low-pass filter during the step evaluation, should give a better result. The analysis with the Modified Method A is presented in TABLE IV. The absolute accuracy error in the measurement of the step length, ranges from 0.32% to 3.33% with a mean value 2.17%. The error is less that in the original Method A. The anthropometric distance is evaluated with C coefficient of TABLE I. With the Method A (TABLE V), the absolute accuracy error in the walking speed ranges from 0.37% to 10.77% with a mean value 5.55%. The stride and step frequencies measured are presented in the same table. With the modified method, A (TABLE VI), the absolute accuracy error in the walking speed ranges from 0.52% to 3.31% with a mean value 1.70%.

TABLE III. METHOD A: REAL AND CALCULATED STEPS AND DISTANCES VALUES FOR THE SIX TESTS. § ONE STEP IS MISSED.

	Steps	Step Length [cm]	Step Length Accuracy Error %		Distance [m]	Distance Length Accuracy Error %	
Test	Real = 52	Real = 60	Relative	Absolute	Real = 31.20	Rel.	Abs.
1	52	59.8 ± 7.6	-0.37%	0.37%	31.08	-0.37	0.37
2	52	56.3 ± 8.4	-6.15%	6.15%	29.28	-6.15	6.15
3	52	57.8 ± 8.2	-3.64%	3.64%	30.07	-3.64	3.64
4	51	57.6 ± 6.9	-4.06%	4.06%	29.36 §	-5.91	5.91
5	52	53.5 ± 8.3	-10.78%	10.78%	27.84	-10.8	10.8
6	51	57.2 ± 8.1	-4.65%	4.65%	29.18 §	-6.48	6.48
mean	52	57.0	-4.94%	4.94%	29,50	-5,22	5,22
min	51	53.5	-10.78%	0.37%	27,84	-10,78	0,37
max	52	59.8	-0.37%	10.78%	31,08	-0,37	10,78

TABLE IV. MODIFIED METHOD A. THE DATA ARE REAL AND CALCULATED STEPS AND DISTANCES VALUES FOR THE SIX TESTS. FOR * THE DISTANCE BY MODEL IS CORRECTED FOR THE ONE STEP MISSED.

	Step Length [cm]		Distance [m] a) Real 31.20 b) Anthropometric 32.21	Step Length Accuracy Error %	
Test	Real	Model	c) Model	Relative	Absolute
1	60.00	61.49	31.98	2.49	2.49
2	60.00	58.00	30.16	-3.33	3.33
3	60.00	59.81	31.10	-0.32	0.32
4	60.00	58.08	30.78 *	-3.21	3.21
5	60.00	58.95	30.66	-1.75	1.75
6	60.00	58.84	31.17 *	-1.94	1.94
mean	60.00	59.20	30.78	-1.34	2.17
min	60.00	58.00	30.16	-3.33	0.32
max	60.00	61.49	31.98	2.49	3.33

The stride and step frequencies measured are presented in the same table. The error is less that in the original Method A. In the Method A (TABLE VII), the absolute error in the COM amplitude versus expected ranges from 1.04% to 18.92% with a mean value 11.05%. In the Modified Method A (TABLE VIII), the absolute error in the COM amplitude versus expected ranges from 0.18% to 5.92% with a mean value 3.88%. The Modified Method A is good: all steps are detected, except in some tests where one is missed, and the accuracy error is acceptable. We have the need for a development of this method for a better detection of all steps and for a more flexible tool to analyze data: the Method B.

TABLE V. METHOD A. TIMES, VELOCITIES, STRIDES, STEPS AND ACCURACY VELOCITY ERROR ARE PRESENTED FOR THE SIX TESTS. § FAILURE RECOGNITION OF ONE STEP.

	Time [s]	Speed [m/s]		Accuracy Speed Error %		Frequency	
Test		Real	Model	Rel.	Abs.	[Stride/s]	[Step/s]
1	32.68	0.9547	0.9512	-0.37	0.37	0.80	1.59
2	31.49	0.9907	0.9298	-6.15	6.15	0.83	1.65
3	30.56	1.0209	0.9837	-3.64	3.64	0.85	1.70
4	31.06	1.0044	0.9451 §	-5.90	5.90	0.80	1.64
5	31.95	0.9764	0.8712	-10.77	10.77	0.81	1.63
6	29.84	1.0455	0.9777 §	-6.48	6.48	0.84	1.71
mean	31.26	0.9988	0.9431	-5.55	5.55	0.82	1.65
min	29.84	0.9547	0.8712	-10.77	0.37	0.80	1.59
max	32.68	1.0455	0.9837	-0.37	10.77	0.85	1.71

TABLE VI. MODIFIED METHOD A. TIMES, VELOCITIES, STRIDES, STEPS AND ACCURACY VELOCITY ERROR ARE PRESENTED FOR THE SIX TESTS.

	Time [s]	Speed [m/s]		Accuracy Speed Error %		Frequency	
Test		Real	Model	Rel.	Abs.	[Stride/s]	[Step/s]
1	33.77	0.9240	0.9546	3.31	3.31	0.77	1.54
2	32.78	0.9518	0.9331	-1.96	1.96	0.79	1.59
3	32.35	0.9644	0.9712	0.71	0.71	0.80	1.61
4	32.09	0.9724	0.9475	-2.56	2.56	0.81	1.62
5	33.05	0.9439	0.9390	-0.52	0.52	0.79	1.57
6	31.69	0.9846	0.9732	-1.16	1.16	0.82	1.64
mean	32.62	0.9568	0.9531	-0.36	1.70	0.80	1.60
min	31.69	0.9240	0.9331	-2.56	0.52	0.77	1.54
max	33.77	0.9846	0.9732	3.31	3.31	0.82	1.64

TABLE VII. METHOD A. EVALUATION OF THE COM AMPLITUDE DURING THE SIX TESTS.

	COM Amplitude [cm]				Mean Versus Expected Error %	
Test	Min	Max	Mean	Expected	Relative	Absolute
1	1.50	6.00	4.86	4.81	1.04	1.04
2	0.82	6.00	4.32	4.81	-10.19	10.19
3	1.25	6.00	4.55	4.81	-5.41	5.41
4	0.91	6.00	4.49	4.81	-6.65	6.65
5	0.39	6.00	3.90	4.81	-18.92	18.92
6	0.61	6.00	4.46	4.81	-7.28	7.28
mean	0.91	6.00	4.30	4.81	-10.71	11.05
min	0.39	6.00	3.09	4.81	-18.92	1.04
max	1.50	6.00	4.86	4.81	1.04	18.92

TABLE VIII. MODIFIED METHOD A. EVALUATION OF THE COM AMPLITUDE DURING THE SIX TESTS.

	COM Amplitude [cm]				Mean Versus Expected Error %	
Test	Min	Max	Mean	Expected	Relative	Absolute
1	3.35	6.00	5.09	4.81	5.92	5.92
2	2.04	6.00	4.55	4.81	-5.39	5.39
3	2.50	6.00	4.82	4.81	0.18	0.18
4	2.57	6.00	4.54	4.81	-5.63	5.63
5	3.67	6.00	4.66	4.81	-3.10	3.10
6	2.84	6.00	4.66	4.81	-3.03	3.03
mean	2.83	6.00	4.72	4.81	-1.84	3.88
min	2.04	6.00	4.54	4.81	-5.63	0.18
max	3.67	6.00	5.09	4.81	5.92	5.92

3.2 The Data Analysis with Method B

The Method B (TABLE XI) detected all 52 steps in every test. We do not yet used the correction for steady state but if we did, the results will be better. By using Mode 5, the absolute accuracy error in the measurement of the walking total path, as well as the step length (error is the same), ranges from 0.27% to 6.31% with a mean value 3.00%. With this modality, the relative values are both positives and negatives.

TABLE IX. METHOD B. MODE 5. REAL AND CALCULATED STEPS AND DISTANCES VALUES FOR THE SIX TESTS. THE METHOD B WITH MODE 5 IS USED FOR DATA PROCESSING.

	Steps	Step Length [cm]	Step (and Distance) Length Accuracy Error %		Distance [m]
Test	Real = 52	Real = 60	Relative	Absolute	Real = 31.20
1	52	63.8 ± 5.2	6.31	6.31	33.17
2	52	60.2 ± 7.5	0.27	0.27	31.28
3	52	60.9 ± 7.8	1.52	1.52	31.67
4	52	60.4 ± 6.2	0.71	0.71	31.42
5	52	57.0 ± 7.5	-4.98	4.98	29.65
6	52	62.5 ± 7.5	4.22	4.22	32.52
mean	52	60,8	1,34	3,00	31.62
min	52	57,0	-4,98	0,27	29.65
max	52	63,8	6,31	6,31	33.17

By using Mode 1 (TABLE X), the absolute accuracy error in the measurement of the walking total path, as well as the step length (error is the same), ranges from 1.58% to 6.80% with a mean value 3.72%. The Mode 1 underestimates the step length in every test. We can think of using a special correction factor to correct this underestimation to have more accuracy. This allows the definition of the proper filtering to be adopted by the two processing methods.

TABLE X. METHOD B. MODE 1. REAL AND CALCULATED STEPS AND DISTANCES VALUES FOR THE SIX TESTS.

	Steps	Step Lenght [cm]	Step (and Distance) Length Accuracy Error %		Distance [m]
Test	Real= 52	Real= 60	Relative	Absolute	Real= 31.20
1	52	59.1 ± 3.4	-1.58	1.58%	30.71
2	52	57.1 ± 5.5	-4.84	4.84%	29.69
3	52	58.0 ± 6.4	-3.32	3.32%	30.17
4	52	58.1 ± 4.7	-3.21	3.21%	30.20
5	52	55.9 ± 6.5	-6.80	6.80%	29.08
6	52	58.5 ± 6.1	-2.55	2.55%	30.40
mean	52	57.8	-3.72	3.72%	30.04
min	52	55.9	-6.80	1.58%	29.08
max	52	59.1	-1.58	6.80%	30.71

In the two conditions 1) and 2) the BMI of the subject was very different. In the trial repetitions of the first protocol (six tests of a single healthy subject) and their data processing, is demonstrated that the model produced the expected values with very good accuracy. The use of coefficients C for evaluating S_{mean} (mean anthropometric step length) carry out a good reference, but is not the measured value S by the model. In the Method A, non-all the steps are detected (one step is lost in two tests), but if the filter is changed as in the Method B, the 100% of the 52 steps are detected. In the Modified Method A, the absolute accuracy error in the measurement of the step length, ranges from a minimum 0.32% to a maximum 3.33% with a mean value 2.17%; in the Method B, it ranges from a minimum 0.27% to a maximum 6.31% with a mean value 3.00%. These values do not yet use the correction for the steady state (used by Modified Method A), but if we did, the results should be better, so the equivalent values of the Method A to be compared are the minimum 0.37%, the maximum 10.77% and the mean value 5.55%. The Method B is better and more flexible. The parameters extracted by the model are very complete, but should be compared with the gold standard values using appropriate protocols. Preliminary reliability of both methods is more than good. This first validation is now followed by a protocol application on a wider population of healthy subjects and post stroke patients, to validate also its clinical application. The tests with an electronically controlled treadmill are carried out on healthy subjects following the dedicated protocol. We are also investigating the clinical applicability of the approach on a group of stroke patients undergoing rehabilitation programs. The tests performed are the 10m test at low and high walking speed, the 6MWT and the TUG. A control group of healthy subjects is considered. The expectation is to develop a dedicated tool for supporting diagnosis and rehabilitation. This will also allow for investigation of model sensitivity in detecting the gait parameters improvements.

REFERENCES

[1] W. Tao, T. Liu, R. Zheng and H. Feng, "Gait Analysis Using Wearable Sensors", Sensors (2012) 12: pp. 2255-2283;

[2] A. Muro-de-la-Herran, B. Garcia-Zapirain and A.Mendez-Zorrilla, "Gait Analysis Methods: An Overview of Wearable and Non-Wearable Systems, Highlighting Clinical Applications", Sensors (2014) 14: pp. 3362-3394;

[3] S. Patel, H. Park, P. Bonato, L. Chan and M. Rodgers, "A review of wearable sensors and systems with application in rehabilitation", Journal of NeuroEngineering and Rehabilitation (2012) 9:21;

[4] I. Spulber, E. Papi, Y.-M. Chen, S. Anastasova-Ivanova, J. Bergmann, P. Georgiou and A.H. McGregor, "Development of a wireless multi-functional body sensing platform for smart garment integration", Biomedical Circuits and Systems Conference (BioCAS), 22-24 Oct. 2014 IEEE;

[5] A.Pantelopoulos, N. G. Bourbakis, "A Survey on Wearable Sensor-Based Systems for Health Monitoring and Prognosis", IEEE Transactions on Systems, Man and Cybernetics, Part C (Applications and Reviews), (2010) 40 (1): pp. 1-12;

[6] W. Zijlstra and A. L. Hof, "Assessment of spatio-temporal gait parameters from trunk accelerations during human walking", Gait and Posture (2003) 18: pp. 1-10;

[7] N. A. Capela , E. D. Lemaire and N. Baddour, "Novel algorithm for a smartphone-based 6-minute walk test application: algorithm, application development, and evaluation", Journal of NeuroEngineering and Rehabilitation (2015) 12:19;

[8] M. S. Orendurff, A. D. Segal, G. K. Klute, J. S. Berge, E. S. Rohr and N.J. Kadel, "The effect of walking speed on center of mass displacement", J Rehabil Res Dev, Nov-Dec (2004) 41 (6A): pp. 829-834;

[9] K. G. Holt, J. Hamill and R. O. Andres, "The force-driven harmonic oscillator as a model for human locomotion", Human Movement Science (1990) 9: pp. 55-68;

[10] http://www.sxt-telemed.it/Products/Protheo/Protheo.aspx, last accessed 22 february 2017;

[11] T. Oberg, A. Karsznia and K. Oberg, "Basic gait parameters: reference data for normal subjects, 10-79 years of age", Journal of rehabilitation research and development (1993) 30 (2): pp. 210-223;

[12] S. Al-Obaidi, J. C. Wall, A. Al-Yaqoub, M. Al-Ghanim, "Basic gait parameters: A comparison of reference data for normal subjects 20 to 29 years of age from Kuwait and Scandinavia." Journal of rehabilitation research and development (2003) 40 (4): pp.361-366;

[13] H. Tanaka, K. D. Monahan, D. R. Seals, "Age-predicted maximal heart rate revisited", J Am Coll Cardiol. (2001) 37 (1): pp. 153-156;

[14] E. Billauer, Peak detect using Matlab, http://www.billauer.co.il/peakdet.html, last accessed 22 february 2017;

[15] R. Moe-Nilssen and J. L. Helbostad, "Estimation of gait cycle characteristics by trunk accelerometry", Journal of Biomechanics (2004) 37: pp. 121-126;

[16] T. Jurcevic Lulic and O. Muftic, "Trajectory of the human body mass centre during walking at different speed", DS 30: Proceedings of DESIGN 2002, the 7th International Design Conference, Dubrovnik (2002).

Proposal of a Validation Protocol for Wearable Systems Reliability Assessment

Application to PEGASO sensors' system

Giuseppe Andreoni, Paolo Perego, Marcello Fusca
Design dept. – E4Sport Lab
Politecnico di Milano
Milan, Italy
email: {giuseppe.andreoni, paolo.perego, marcello.fusca}@polimi.it

Martina Caramenti, Fabio Rastelli, Claudio L. Lafortuna
Istituto di Bioimmagini e Fisiologia Molecolare
CNR
Segrate (Milan), Italy
email: {martina.caramenti, fabio.rastelli, claudio.lafortuna}@cnr.it

***Abstract*—This paper describes the proposal of a validation procedure to assess the reliability of wearables systems. According to the two main categories of devices we have defined an integrated experimental protocol composed by two parts: the first one applies to sensorized garments or similar devices (also including belts or patches) while the second part is dedicated to body worn activity trackers like smart bracelets or smart watches, belt-worn devices, sensors embedded in necklace or other garment accessories. This protocol applied a sequence of real life activities and posture changes to be properly identified in value and in time. It has been applied to the case study of the sensors developed and used in the frame of the Pegaso Fit 4 Future EU project. This test has demonstrated the protocol feasibility, applicability, easy to use and easy data processing and its sensitivity. We are adopting this protocol as standard procedure in any further study regarding the reliability assessment of wearable devices.**

Keywords-wearable systems; validation protocol; reliability; accuracy; real life testing.

I. INTRODUCTION

In the last years, a worldwide spreading category of devices used in sport, fitness and also healthcare are wearable systems [6]. Wearable devices are a popular and growing market for monitoring physical activity, sleep, and other behaviors [2]. In particular, for healthcare applications, activity trackers like wrist-worn systems (bracelets) or similar devices are now in progress, but this requires a deep and structured reliability analysis of their performances before their deployment into clinical practice. Several studies are ongoing [2], each using its own protocol. For this reason, and to methodologically have the opportunity to compare homogeneously these technologies, we have decided to study and propose a dedicated standard protocol for reliability assessment here described.

The paper is divided into two sections: Section 2 highlight the experimental protocol underlining the difference between the first part (static test) and the second one (dynamic test). Section 2 continues with the explanation of the wearable sensors tested and closes with reporting the validation procedure and parameters. Section 3 reports the results on both testing procedure and the application of this on specific wearable device. Then Section 4 ends the paper with the conclusion and future development.

II. MATERIALS & METHODS

A. Experimental protocol

The experimental protocol is methodologically based onto a paired video and bio-signal acquisition during different postural tasks in controlled environment and locomotion (walking and running) at different velocities on treadmill (Woodway Inc., USA). The video is recorded by a digital camera with lateral acquisition of subject. The subjects followed a supervised protocol divided in two parts (protocol part I and part II), with the aim of testing the capability of the sensor to recognize body position changes and transitions from rest to activity and vice-versa (protocol part I) and to detect gait and speed of locomotion together with kinematic parameters, during activities on treadmill ranging from sedentary to vigorous intensity physical activity (part II).

The experimental activity is to be conducted in controlled conditions during different postural tasks and locomotion on treadmill with simultaneous video-recording.

A methodologically coherent subject sample, different in age, sex and anthropometry, has to be properly selected.

a) Part I

The first part of the protocol (Part I) is specifically dedicated to assess the capability of the systems to detect posture, its transitions and the related activity level through HR (heart rate). The first part of the protocol (Part I) consists of 10 postural tasks (phase from 1 to 10). More precisely, there is a sequence of postural transitions alternating phases of resting (lying, standing and sitting) and walking/running phases. The overall Part I protocol is described in the following table I.

Contextually, the video of the experiment is recorded by a digital camera with lateral acquisition of the subject.

The assessments indexes are activity recognition, posture recognition and computational time.

a) Part II

Instead, the second part of the protocol (Part II) is specifically designed for assessing the reliability of systems during the progressively increase of activity level. It consists of 10 stages (phase from 1 to 10).

TABLE I. PROTOCOL PART I

Phase no.	Wearable validation protocol – Part I	
	Activity description	*Posture*
1	20 seconds sitting	sitting
2	20 seconds resting with arms at the sides	standing
3	20 seconds walking onto a treadmill	standing
4	30 seconds running onto a treadmill	standing
5	10 seconds sitting	sitting
6	20 seconds lying onto the back	lying (supine)
7	20 seconds lying onto the right side	lying (right)
8	20 seconds lying onto the left side	lying (left)
9	20 seconds lying onto the belly	lying (prone)
10	20 seconds resting with arms at the sides	standing

TABLE II. PROTOCOL PART II

Phase no.	Wearable validation protocol – Part II	
	Activity description	*Posture*
1	Rest sitting for 30 seconds	sitting
2	Rest standing for 30 seconds.	standing
3	Walking at 2 km/h for 30 seconds	walking
4	Walking at 3 km/h for 30 seconds	walking
5	Walking at 4 km/h for 30 seconds	walking
6	Walking at 5 km/h for 30 seconds	walking
7	Walking at 6 km/h for 30 seconds	walking
8	Running at 6 km/h for 30 seconds	running
9	Running at 7 km/h for 30 seconds	running
10	Running at 8 km/h for 30 seconds	running
11	Running at 9 km/h for 30 seconds	running

More precisely, it consisted of 120s of resting (standing and sitting), 150s of walking and 120s of running. The subject carries out the following experimental procedure on a treadmill: from standing to running at 9 km/m then stop and recovery. The overall Part II protocol is described in table II.

Also in this case, contextually, the video of the experiment is recorded by a digital camera with lateral acquisition of the subject. The video of the subject's trial allows for the retrieval of the true values of the computed parameters (activity, steps, cadence, etc.) along with the acquisition.

B. *Wearable system*

To apply the protocol on a specific healthcare application we decided to adopt the sensors developed in the EU project PEGASO Fit 4 Future [3]. In the PEGASO project, the sensors system is composed by 2 elements (Figure 1):

- a sensorized t-shirt (male version) or bra (female version) including a pair of textile electrodes and mounting at the chest level a device named WES capable of recording and/or transmitting 1 ECG lead and 3D accelerations of the trunk.
- an activity tracker consisting in a smart bracelet named WWAT and integrating a 3D accelerometer to monito human kinetics. The WWAT embedded algorithm computes the steps number, the activity (resting, walking, cycling, running, swimming and sleeping) through. Energy expenditure is estimated too.

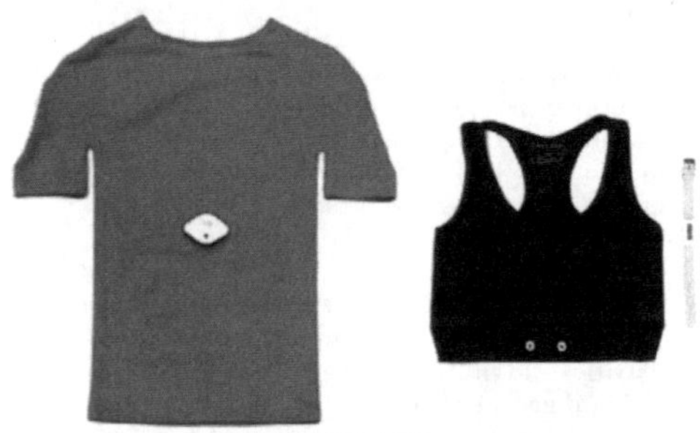

Figure 1: The PEGASO sensor system

C. *Validation parameters, citeria and indexes*

According to the device differences, we have identified to different analyses that can be conducted on corresponding data.

a) Part I and Part II, devices with ECG/Heart Rate measurement

All the signals have to be processed to compute the following three values for each phase:

- Activity recognition (0=N, 1=Y),
- Posture recognition (0=N, 1=Y),
- Computational time (0 = time > 3 sec for activity and posture recognition, 1 = time <= 3 sec for activity and posture recognition).

The test is passed if the overall score is equal or greater than 90% during the complete test.

All the signals have to be processed to compute the beat-to-beat HR with standard algorithms. The values of HR recognition are 0=N and 1=Y. METs (Metabolic Equivalent of Task) estimated from sensor, are compared with values obtained with ACSM (American College of Sports Medicine) metabolic equations for walking/running [4]. The test is passed if the overall score is equal or greater than 90% during the whole test.

Analysis of video-recording, is carried out through Advene, an open source software (Advene, Lyon, France) [5], providing a model and a format to share annotations about digital video documents, as well as tools to edit and visualize

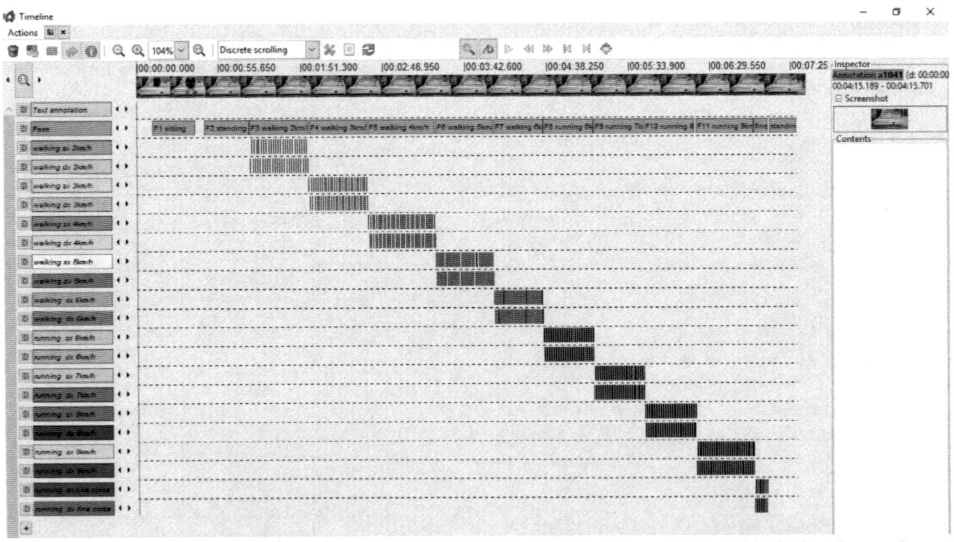

Figure 2. Example of the video-recording data visualization and values together with relative annotations in the Advene software.

the hyper videos generated from both the annotations and the audiovisual documents [7]. An example of the processing procedure of the video-recordings is shown in Figure 2.

b) Part I and Part II, devices with human kinetics measurement only

All the signals have to be processed to compute the following three values for each phase:

- Activity recognition (0=N, 1=Y),
- Posture recognition (0=N, 1=Y),
- Computational time (0 if time > 3 sec for activity and posture recognition, 1 if time <= 3 sec for activity and posture recognition).

This general test is passed if the overall score is equal or greater than 90% during the complete test.

More in detail data are processed in order to verify the correct step identification, count and categorization according to the different activities and/or speed. Also in this case, the test is passed if the overall score is equal or greater than 90% during the complete test. The analysis of video-recording, through Advene, provides the true and reference values.

The experimental activity was conducted in the Laboratory of Biomechanics "Franco Saibene" (Istituto di Bioimmagini e Fisiologia Molecolare del CNR, Segrate, Milan, Italy) in controlled conditions during different postural tasks and locomotion on treadmill with simultaneous video-recording.

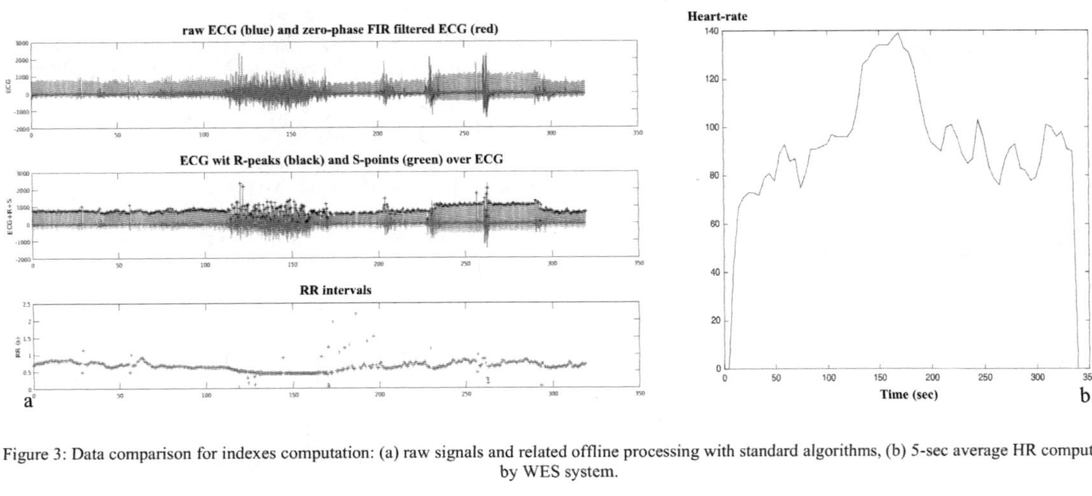

Figure 3: Data comparison for indexes computation: (a) raw signals and related offline processing with standard algorithms, (b) 5-sec average HR computed by WES system.

10 subjects took part in the Part I and Part II experiments for the test of the sensorized garment and the WES device. Their anthropometric data are reported in Table III. Three subjects applied the Part II of the protocol to assess its proper feasibility and reliability on the WWAT device (smart bracelet): they are the subjects no. 1, 8, and 9 in the same Table III.

The study was approved by the competent Institutional Review Board. Subjects were properly informed about testing procedures, personal data treating, and aims of the research, and they provided informed written consent before participation.

III. Results

The results of this study can be divided into two sections: the first one relates to the assessment of the method itself in terms of its sensitivity, feasibility and robustness; the second set of results is about the reliability assessment of the sensors used in the PEGASO project.

TABLE III. Data of subjects participating in wearable device validation

Subj. no.	Subjects recruited in the validation protocol					
	sex	*Age (yrs)*	*heigth (m)*	*weigth (kg)*	*garment*	*BMI*
1	M	47	1,860	105,2	t-shirt	30,41
2	F	33	1,555	56,2	bra	23,24
3	F	33	1,665	52,9	bra	19,08
4	F	34	1,570	53	bra	21,50
5	F	33	1,685	68,9	bra	24,27
6	M	53	1,730	68,8	t-shirt	22,99
7	M	33	1,780	85,6	t-shirt	27,02
8	F	29	1,710	64,5	bra	22,06
9	M	51	1,815	79,6	t-shirt	24,16
10	M	63	1,745	68,7	t-shirt	22,56

A. Methodological results

The protocol we designed try to answer in an integrated test to all the parameters a wearable device is usually designed for. It is robust and sensitive to raw and processed temporal, kinematic and cardiac data. The validation indexed and intuitive and easy to be computed, as well as they are representative of the needed accuracy assessment.

The application case study demonstrated its good applicability and repeatability in a standard sample size of 10 subjects.

B. Results of the protocol on the reliability of the PEGASO Sensors system

The WES sensors have a mean accuracy of 99%±2% in the first part and 93%±7% in the second part; moreover 7 subjects on 10 had an accuracy of 100% in the detection of steps. These data show that WES sensors are reliable and accurate in their measurements, appropriately for the applications they are design to be used for.

TABLE IV. Data of subjects participating in wearable device validation

Percentual Accuracy			
Condition	***Speed (km/h)***	***Total***	***walk/run***
walk	2	65,85%	45,05%
walk	3	18,00%	9,36%
walk	4	10,54%	3,45%
walk	5	12,08%	3,54%
walk	6	12,98%	3,54%
run	6	9,23%	0,97%
run	7	9,26%	0,88%
run	8	9,23%	0,80%
run	9	8,55%	7,26%
	Average	17,30%	8,32%
	St.Dev	0,18	0,14

The WWAT bracelets presented the results shown in table IV. Algorithms refinement for low speed step detection was identified as needed.

IV. Conclusions

This study aims at setting up a standard protocol for wearable sensor validation before their introduction in research projects. This has particular relevance when clinical applications of wearable sensors have been carried out. The protocol we have defined is up-to-date with current technologies and their applications. The results showed both methodologically and for its application a good performance in usability and outcomes. Further application will confirm these findings and other developments could be done for specific detail of garments or activity trackers depending on peculiar applications.

Acknowledgment

This work has been partially funded by the European Commission: FP7-ICT-2013.5.1—Grant Agreement n″ 610727. The Authors would like to thank all the partners of the PEGASO consortium for their proactive collaboration in the project.

References

[1] K. R. Evenson, M. M. Goto, and R. D. Furberg, "Systematic review of the validity and reliability of consumer-wearable activity trackers," International Journal of Behavioral Nutrition and Physical Activity. Vol. 12:159, pp.68–73, 2015.

[2] P. Perego, M. Tarabini, M. Bocciolone, and G. Andreoni, "SMARTA: Smart Ambiente and Wearable Home Monitoring for Elderly," Internet of Things. IoT Infrastructures, Second International Summit, IoT 360° 2015, Rome, Italy, October 27-

29, 2015. Revised Selected Papers, Part I, Lecture Notes of the Institute for Computer Sciences, Social Informatics and Telecommunications Engineering, vol. 169, pp 502-507, 18 November 2016.

[3] M. R. Guarneri and G. Andreoni, "PEGASO Fit for Future," In: Antonello Rizzi and Antonio Pedotti (eds): European Project Space on Computational Intelligence, Knowledge Discovery and Systems Engineering for Health and Sports, SCITEPRESS Publications. Lda, pp. 77-96, 2015.

[4] ACSM certification, "Metabolic Equations for Gross VO2 in Metric Units," https://certification.acsm.org/metabolic-calcs, last access 27 february 2017.

[5] Advene, "Advene Software, " http://liris.cnrs.fr/advene/, last access 27 february 2017.

[6] R. Nensli, E. Gunnarson, and T. Gundersenm, "Wearable medical devices for tele-home healthcare," Proceedings of The 26th Annual International Conference of the IEEE Engineering in Medicine and Biology Society, 2004, pp. 5384-5387.

[7] O. Aubert and Y. Prié, "Advene: active reading O. hypervideo," Proceedings of ACM Hypertext'05, Sep 2005, pp. 235-244.

FrailSafe eCRF

Clinical Data collection tool

Luca Bianconi, Cristiana Degano, Matteo Toma, Fabio Podda
Research Business Unit
Gruppo SIGLA S.r.l.
Genova, Italy
e-mail: frailsafe@grupposigla.it

Abstract—**In the context of European funded FrailSafe project, to enhance the clinical data collection processes in geographically distributed trials centers, an eCRF software platform has been designed and implemented. The most important features are synthetically described.**

Keywords: eCRF; Cloud; eHealth; Monitoring; Data collection.

I. Introduction

The Case Report Form (CRF) may be defined as "a printed, optical or electronic document designed to record all of the protocol" [1]. Extending this definition, a CRF could be described as a tool designed to collect data about a patient state during a clinical trial, e.g., the patient information, the measurements and others. It represents a significant instrument for the performance of a clinical trial and its efficiency could positively affect the success of a medical research.

CRF is designed for improving the collection of data in compliance with a research protocol and with regulatory requirements. Moreover, the researchers should be able to test the hypothesis or answer the trial related questions. A CRF represents suitably the essential contents of the defined research and study protocol.

A flexible eCRF has been implemented for FrailSafe project to provide the partners devoted to clinical research of such features.

II. FrailSafe eCRF

In the current global scenario of FrailSafe project, the development of an electronic CRF (also known as *eCRF*) has been preferred over a usual paper CRF for the following benefits:

- Eliminate unnecessary duplication of data.
- Reduce the possibility for transcription errors.
- Encourage entering source data during a subject's visit, where appropriate.
- Eliminate transcription of source data prior to entry into an eCRF.
- Facilitate remote monitoring of data.
- Promote real-time access for data review.
- Facilitate the collection of accurate and complete data.
- Possibility to have auto generated data (test results, calculate formulas, etc.).
- Easier data export and data analysis.

As mentioned above, the eCRF aims to support the clinical researchers during the trials by providing them a web application (appearing in this case as a plain website) through which tracking and managing the process of visiting the participants and enhancing the related data collection, as depicted in Figure 1.

With the objective of supporting the trial processes in mind some functional features are needed.

These are:

- Users Management.
- Participants Management.
- Trials Management.
- Devices Management.
- Data Collection.

For enhancing the quality, the reliability, the usability and the security to reach the common goal, some other aspects of the application have been taken into consideration while designing the software.

They are detailed in the following sub-paragraphs.

A. Cross-platform and Multi-device

The eCRF is a web application. This means it behaves and looks like a common website and that it can be used through any simple web browser (e.g., Chrome, Firefox, Internet Explorer, Safari and Opera); thanks to that, it is immediately available on every device providing a web browser and an Internet connection (laptop, smartphone, tablet, etc.), not needing any further technical requirements. A mobile-first approach has been followed in the design of the UIs.

B. Authentication and Encryption of Data

Security is guaranteed through the usage of authentication means, i.e., a username and a password are needed for accessing to the application and its resources – and through data encryption. Data is encrypted before being stored on database and, in communication, a standard protocol for secure communication over a computer network is adopted to protect the privacy and the integrity of exchanged data.

C. *Data quality*

In the eCRF, the clinical staff collects the patient information during the trial in the same ways and means as if doing it on paper - that is through questionnaires, forms, etc. – but displayed instead on a device screen. During the data addition or before saving, the system checks the coherence of the data and in case something is not correct it helps the user with feedbacks. For example, the participant information cannot be inserted in the system if all the mandatory fields are not filled or it is not allowed to type characters in a field that is a measurement value, etc.

D. *Multilanguage*

The eCRF system, for the FrailSafe project, is used in three different countries (France, Greece, Cyprus); so the system can support the different languages to help the clinical staff and the participants to avoid translation errors or misunderstandings during the trials execution.

III. Cloud Infrastructure

For guaranteeing a perpetual availability of the system to the clinical staff, the eCRF is hosted on remote servers. On these latter the collected data, thus, are stored securely and can be used by the system and by the researchers to succeed in the medical research carried on by FrailSafe.

Moreover, the efficiency and the security of the system rely on a number of Cloud services and features.

IV. Conclusions

In conclusion, this web application is designed and implemented for assessing correctly the objectives of the project and of the related medical research. Moreover, it is worthy to add that the designing and the implementation of the eCRF has been an important experience for those having contributed to its design and implementation. It proved to be a good example of interdisciplinary and collaborative production of a technological tool. It has seen the close participation of clinical and ICT partners, from different countries, making them willingly and passionately working together to reach a common goal, i.e., implementing the clinical trial protocol by the mean of an ICT instrument – showing evidently the benefits of sharing different backgrounds and skills.

Acknowledgment

Gruppo SIGLA Research Team thanks the FrailSafe Consortium for the commitment and energy in the collaboration whose, the eCRF is the successful outcome.

FrailSafe is funded by the European Commission under H2020 - PHC-21-2015 - Advancing active and healthy ageing with ICT: Early risk detection and intervention. Project ID: 690140.

For further details, refer to [2].

Figure 1. Main Clinical Staff dashboard

References

[1] Guideline for Good Clinical Practice, ICH Harmonised Tripartite Guideline, 1996.
[2] www.frailsafe-project.eu

International Academy, Research, and Industry Association (IARIA)
412 Derby Way
Wilmington, DE 19810

ISBN 978-1-5108-3813-0

CURRAN ASSOCIATES INC.
proceedings.com